VoIP FOR DUMMIES®

by Timothy Kelly

Foreword by Don Peterson
Chairman and Chief Executive Officer, Avaya Inc.

Wiley Publishing, Inc.

VoIP For Dummies®

Published by
Wiley Publishing, Inc.
111 River Street
Hoboken, NJ 07030-5774

www.wiley.com

Published by Wiley Publishing, Inc., Indianapolis, Indiana

Published simultaneously in Canada

Library of Congress Control Number: 2005923780

ISBN-13: 978-0-7645-8843-3

ISBN-10: 0-7645-8843-5

Manufactured in the United States of America

10 9 8 7 6 5 4 3

1B/RX/QY/QV/IN

About the Author

Timothy Kelly is an Information Systems technology professional with more than twenty-five years of experience. His background includes the design of many telecommunications network enterprises, from small simple networks that support a single building location to large multilocation networks running integrated data, voice, and videoconferencing applications.

From 1992 until 2002, Tim was principal consultant for Network Technology Services, a Pittsburgh-based company. He has completed network design engagements for countless organizations, including Alcoa, Blue Cross, Mercy Health System, Mine Safety Health Administration, the U. S. Navy, South Hills Health System, Westinghouse Telecommunications, ARBROS Communications, The Community Builders, and Lucent Technologies. Kelly is a certified ORACLE DBA Master and Network+ Professional.

Tim is an honors graduate of Duquesne University. He went on to complete the MSIS and post-graduate certificate in Telecommunications at the University of Pittsburgh. His terminal degree is a Doctor of Science in Information Systems from Robert Morris University. His research focus was the effects that converging technology networks have on organizations and people, an area in which he is well published and has made numerous presentations before academic and corporate bodies.

Tim Kelly is author of *Bits & Bytes Y2K & Beyond* and is well known for his consults and media appearances during the years and final months preceding the year 2000. He was dubbed a "calming influence" on the Y2K scare by the Pittsburgh media.

From 1983 to 2004, he taught Information Systems Technology courses for local Pittsburgh schools, including Duquesne University, Indiana University of Pennsylvania, and Robert Morris University. In 2003, with the help of former associates, he started the National Center for Converging Technology Research, an organization dedicated to helping other organizations understand how best to apply converging technologies such as VoIP in their business environments. In Fall 2004, he began teaching full-time for the University of North Carolina at ECSU.

Tim Kelly will be co-authoring a VoIP solutions book that defines the latest convergence options for running data, voice, and video applications — the "triple play." The book will provide current coverage on the latest wireless forms of networking. The effect on business of WiMax and other fixed-wireless alternatives will be treated. Tim believes the solution to the triple play model lies with resolving the dilemma of inadequate bandwidth and that VoIP over WiMax and WiFi show how close we are to cracking this nut. The next few years for VoIP should be really exciting.

Dedication

To my primary passion source, my heart and soul, my Tushka; and our four children: Laural, Christal, Gabe, and Matt. Each beat of my heart has four distinct iterations.

Author's Acknowledgments

I would like to thank all of my friends at Robert Morris University. They gave me a great deal to think about, chief among which was the need to put VoIP convergence into a frame that the average reader could understand. In our discussions about my ideas, I would constantly hear "think Dummies." With the dramatic changes in the VoIP convergence marketplace in 2004, I knew the time was right to not only think Dummies but to also write Dummies.

I would like to express a truly heartfelt thanks to Greg Croy at Wiley Publishing. Greg believed in my ideas and supported me through the entire process. It is not easy to become a *For Dummies* author, but the guidance from Greg enriched the process while making it possible and enjoyable. I also want to thank Leah Cameron, who conducted the first nuts-and-bolts review of my work. Her feedback was invaluable. I want to thank Nancy Stevenson for her help in finalizing the Table of Contents.

I want to thank Allen Wyatt and Dave Tegtmeier, my preproduction editorial team. Assembling a group of this magnitude was no small challenge as professionals of this caliber are always in demand. But they all found the time to take on *VoIP For Dummies.* Words cannot express my appreciation.

I want to thank Chuck Mance, a friend of mine who lent a hand with drafting Chapter 14. Chuck is an experienced, competent IT professional. I greatly appreciate his contributions.

I also want to thank the other people who engaged my ideas about VoIP in varying degrees: Steve Phillips, Rich Krauland, all my friends at Avaya, Cisco Systems, Verizon Communications, Matt Kelly, Greg Chmiel, and all of my students and clients.

To my wife Patty (Tushka), who proofed many initial drafts but, more importantly, also helped me get to church on time and provided emotional support throughout the process.

Last but far from least, I want to thank my mother, Mary (Andreiczyk) Kelly, who gave me faith, love, and perseverance. Mom turns 80 in a few short months.

Publisher's Acknowledgments

We're proud of this book; please send us your comments through our online registration form located at `www.dummies.com/register/`.

Some of the people who helped bring this book to market include the following:

Acquisitions, Editorial, and Media Development

Project Editor: Susan Pink

Acquisitions Editor: Greg Croy

Technical Editor: Dave Tegtmeier

Editorial Manager: Carol Sheehan

Media Development Supervisor: Richard Graves

Editorial Assistant: Amanda Foxworth

Cartoons: Rich Tennant (`www.the5thwave.com`)

Composition Services

Project Coordinator: Maridee Ennis

Layout and Graphics: Jonelle Burns, Denny Hager, Stephanie D. Jumper, Heather Ryan

Proofreaders: Leeann Harney, Jessica Kramer, Linda Morris, Dwight Ramsey

Indexer: Joan Griffitts

Special Help

Allen Wyatt

Publishing and Editorial for Technology Dummies

Richard Swadley, Vice President and Executive Group Publisher

Andy Cummings, Vice President and Publisher

Mary Bednarek, Executive Acquisitions Director

Mary C. Corder, Editorial Director

Publishing for Consumer Dummies

Diane Graves Steele, Vice President and Publisher

Joyce Pepple, Acquisitions Director

Composition Services

Gerry Fahey, Vice President of Production Services

Debbie Stailey, Director of Composition Services

Contents at a Glance

Table of Contents

Foreword

Communications is the heart of your business, and voice over IP has the capability to strengthen that heart and thereby strengthen your business. VoIP is not just another form of connectivity. Yes, it combines the intimacy of voice with the power of data, but it is more than voice over the Internet or voice over your data network. It enables the merging of voice and data applications in ways that liberate business processes. VoIP extends voice communications to anyone, anywhere, over any device — it is the fundamental building block of intelligent communications. It offers businesses the benefits of significant cost savings, increased revenue, and better customer service. It puts communications at the core of the business, enabling faster decisions, revitalized business processes, and new business models.

This year is a pivotal one in electronic communications. With customer confidence growing, IP is now preferred over traditional phone systems. With VoIP becoming mainstream, the adoption rate is accelerating.

Voice over IP is no longer a wait-and-see decision. It's happening right now. You can't afford to limit your communications options or neglect the role that it can play in business performance. But maximizing success in switching to a VoIP system requires top-notch planning, design, implementation, and management. To help you get started and understand the fundamentals, Tim Kelly has written a fine book, *VoIP For Dummies*. This book lifts any confusion you may have about the subject and clearly identifies the many benefits of VoIP for businesses. This book is your portal to understanding how VoIP can make your business stronger by making your communications systems stronger. The results will be people more productive, processes more efficient, and customers more loyal.

Don Peterson

Chairman and Chief Executive Officer, Avaya Inc.

Introduction

VoIP (pronounced *voyp*) is the name of a new communications technology that changes the meaning of the phrase *telephone call.* VoIP stands for *voice over Internet protocol,* and it means "voice transmitted over a computer network."

Internet protocol (IP) networking is supported by all sorts of networks: corporate, private, public, cable, and even wireless networks. Don't be fooled by the "Internet" part of the acronym. VoIP runs over any type of network. Currently, in the corporate sector, the private dedicated network option is the preferred type. For the telecommuter or home user, the hands-down favorite is broadband.

You may be wondering what all this means in terms of your actual telephone. This is the really cool part: You can access your account on the VoIP network by a desktop telephone, a wireless IP phone (similar to a cell phone), or the soft screen dialpad of your laptop or desktop computer.

With VoIP, you can literally pick up your things and move to another location, within your office building or around the world, without having to forward your calls to a new telephone. VoIP's entirely portable!

What's more, you can access the Web from your IP phone, enabling you to get important (or not so important) announcements and e-mail on the go. It's like having a pocket PC and a cell phone rolled into one, specifically designed for *your* network.

As you can imagine, VoIP is a win-win for everyone. The added flexibility and quicker response times translate into greater customer satisfaction and increased productivity throughout your organization.

About This Book

VoIP For Dummies is written for anyone who wants to reduce or eliminate their toll charges while upgrading the level of computer networking services and calling features they receive. Here you discover not only what VoIP is but how you can implement it in your company or home. (You'll even find out whether VoIP makes a lot of sense for your situation.)

VoIP has particular appeal to those who want to use their computer network to carry their telephone calls, thereby saving the expense of running different networks for each.

If you're a consumer running broadband Internet services and you have significant toll charges each month, you should look into VoIP to make your toll calls. With VoIP running on your broadband line, you can save money each month by reducing your toll costs while still maintaining your traditional telephone service for local calls.

If you're a manager who needs to decide about support or recommend whether to make the switch to VoIP, or if you're an IT person looking to help your boss make an informed decision about integrated networking, this book provides an excellent place for you to begin.

I explain how VoIP works and how it compares to telecommunications technology that was previously considered irreplaceable. By the time you finish Part III, you'll see why many businesses throughout the world and consumers in the United States have turned to VoIP and integrated networking as their main system for data, voice, and video.

Conventions Used in This Book

To help you navigate through this book, I use the following conventions:

- *Italic* is used to highlight new words or terms that are defined.
- **Boldfaced** text is used for chapter titles, subtitles, and to indicate keywords in bulleted lists.
- `Monofont` is used for Web addresses.
- Sidebars, unlike the rest of the content, are shaded in gray.

What You're Not to Read

Whether you are a consumer or a corporate user, you don't have to read this book from cover to cover to find out how VoIP can benefit you or your company. You may miss some really interesting stuff, but if you're interested in knowing just the fundamentals of IP telephony and VoIP, you can get that information by reading just Chapters 1 and 2. These two chapters cover VoIP basics and introduce you to how you can make VoIP work for you.

If you're unfamiliar with how traditional telephone companies bill their customers (that's you!), Chapter 3 enlightens you with this information. (Before reading this chapter, you need to promise that you won't yank the phone cords out of the wall when you discover how much you are really paying — talk *isn't* cheap!)

If you're thinking of putting VoIP in your home or even in your home office, or you already have done so, you may be interested in gaining more information about VoIP fundamentals in Part I and then reading Chapter 6, where I describe how to put broadband VoIP to work in your home. If you're using VoIP from home to connect to your company's virtual private network (VPN), you'll also want to look at Chapter 9.

Information technology professionals working in the corporate world, and the people that manage them, will be more interested in Chapters 4 through 7 than any other section of the book. These chapters cover all the VoIP network types used in the corporate sector.

If you just want to define the type of telephone your company is currently using or may use with VoIP, check out Chapter 10. If you need to understand the traditional non-VoIP telephony system models that a company must have to even begin to look at VoIP, see Chapter 11.

If you want to move your company toward a VoIP telephony system model, you need to know how to make it work from a financial perspective; Chapters 12 and 13 can help with case studies and cost figures. These chapters detail how a multilocation company and a smaller single-location company can transform their monthly telephony system finances using a VoIP network. Chapter 14 details other factors that apply to evaluating a move to VoIP for any size network.

Feel free to read this book from cover to cover or just dip into whatever part or section best suits your needs. You can then return to the rest of the book when you have more time to enjoy the read.

Foolish Assumptions

As I wrote *VoIP For Dummies,* I made some assumptions about you and what you might already know about traditional telephony services in contrast to VoIP telephony. Here are those assumptions:

- You probably have trouble understanding your monthly telephone bills and don't realize that their long-distance is divided into four billable service categories.
- You rarely consider that there is a cost for the line (access line) and a cost for the usage on that line.
- You might be thinking that VoIP is a new way of doing telephony but, from what you've heard, it works only over the Internet.
- You may know the basics of computer networking and VoIP, but you want to gain advanced knowledge, like using your computer and your older POTS phone simultaneously with your new VoIP service.
- You've heard about all the new and exciting features that come with VoIP at no additional cost.
- You heard (incorrectly) that 911 and E911 do not work with VoIP, not knowing that VoIP principles are the technology that underlies E911.
- You've heard that VoIP can save the consumer or the company lots of money.
- You may want to protect your company's telephony systems investment while figuring out a way to bring VoIP in because you know it will save the company big bucks.

How This Book Is Organized

Each part of this book focuses on a different aspect of VoIP, as described in the following sections. VoIP is a technology that challenges all your preconceptions about telephony and networking.

Part I: VoIP Basics

Part I introduces you to the basics of VoIP. You get the rundown on essential terms and the general workings of the technology. This part describes the basics of IP telephony and how VoIP calls get packetized and carried over external networks. Access services and the lines they run on are defined. You find out how traditional telephony models can become cash cows for the carriers. TCP/IP, the number one network design model, is introduced as the underlying design for VoIP networks.

Part II: Taking VoIP to Your Network

In Part II, you discover how networks connect to each other. From the public telephone network to the global Internet and all the network types in between, you'll find out what your networking options are.

The Internet is only one network option for VoIP (it also runs on all the other network types that drive industry). Network types include broadband networking, which exists mainly as a consumer option for VoIP. Other types covered in Part II are switched, dedicated, and wireless networks. There is no shortage of network types to run VoIP on.

To help set VoIP in a network context, Part II compares the transport lines (where applicable) and services available on each network type. Also covered are bandwidth options and quality of service. With these options, companies can support not only VoIP but their data and videoconferencing needs for all their locations.

Consumers are also treated to illustrated coverage on broadband networking options. You can run VoIP out of your home and receive ideal bandwidth options that support not just VoIP but your computer data — and even video.

When it comes to VoIP, all the network options in the world would be of little value if you couldn't actually talk on the phone! For that reason, Chapter 10 outlines the major VoIP-enabled telephone types: VoIP hard phone, VoIP soft phone, and VoIP wireless phone. It also covers the traditional telephone types that can be used in a VoIP network.

Part III: Making the Move to VoIP

The reasons to switch to VoIP are countless, depending on how far you want to project the future of the marketplace. Part III starts in Chapter 11 by describing the "final four" telephone system models. These are the traditional systems used by consumers and corporate customers. If you're not yet on VoIP, you must be running with one or more of the final four options.

Through real-world case studies, Part III provides guidance for both single-location and multilocation companies, covering the total cost factors and then applying a VoIP solution that significantly reduces the cost of a VoIP conversion while enhancing productivity.

Part IV: The Part of Tens

In Part IV, I provide specific content and advice for both corporate and consumer-based prospective VoIP users. This, of course, is accomplished in the time-honored *Dummies* format: the venerated Part of Tens.

If you're a company pondering the move to VoIP, Chapter 15 gives you the top ten reasons why you need to make the move. Consumers find the top ten reasons why they should change in Chapter 16. Chapter 17 dispels the top ten myths about VoIP. Get the truth about these myths here and now.

Finally, Chapter 18 provides a quick overview of the best of the best: the top ten VoIP manufacturers. When you're ready to make the move, you'll know who to go to for support.

Part V: Appendixes

Last, but by no means least, the final section of this book includes two reference items that you will find helpful in making sense of the world of VoIP. The first, Appendix A, provides an overview of the largest VoIP service providers in the world. These are the companies that you can partner with to realize all your VoIP dreams.

The second item is a handy glossary. Confused by a term you encounter while reading the book? Turn to the glossary and your bewilderment will fade into the past. (It's also a great tool for understanding VoIP marketing brochures and white papers.)

Icons Used in This Book

Throughout this book, I occasionally use icons to call attention to material worth noting in a special way. Here is a list of the icons, along with a description of each.

If you see a tip icon, perk up! You're about to find out how to save time, money, or effort. These are the nuggets that, when heeded, can make your life simpler.

This icon indicates information that is probably most interesting to those with a technical bent. If you're responsible for any aspect of the company network or feel comfortable hacking it alone at home on broadband, you'll have no problem breezing through information marked in this manner.

Some points bear repeating, and others bear remembering. When you see this icon, take special note of what you're about to read.

How many times have you heard the phrase *buyer beware*? In paying for traditional telephony and VoIP networks, most concerns revolve around cost and quality of service. When you see this icon, your life won't be in danger, but you will want to pay attention to the "gotcha" that this icon undoubtedly marks.

Where to Go from Here

The most important thing to keep in mind whenever you're exploring a new technology is how it fits into the larger picture. Take a global view. Specifically, always be thinking, "How will this feature increase my company's efficiency?" Or, "How will an integrated network help promote collaboration across my company?" Of course, you may also be wondering how you'll save money with VoIP.

Consider the direction of the telephony industry. The move toward VoIP is happening right here and right now. If you're a consumer, the question is no longer, "Should I get VoIP or broadband services in my home?" Instead, the question is "How do I get these services?"

If you're a decision-maker in your company, you need to strategize how to remain competitive in a constantly changing market. If you're a corporate professional working in a department such as IT, telecommunications, networking, or even finance, you need to research the available technologies so you can make recommendations to your boss and implement, if necessary, a VoIP system. End users need to be prepared to make the switch if their company adopts a VoIP system, or if they get transferred to a new location that already has such a system in place.

This book provides a great place for getting your feet wet, whether you're a consumer, a manager, in charge of the company finances, or an end user. My best advice on where to go from here? Flip the page and keep reading!

Part I
VoIP Basics

In this part . . .

Want to know how VoIP works? You find that information here, discussed in all its glory. Along the way, you also discover a new VoIP terminology, which is essential if you want to make sense of this brave new world. You also get a glimpse at the new and exciting features that are part and parcel with VoIP.

In short, this part reveals the nuts 'n' bolts of VoIP and invites you to a whole new world of networking. Cool, huh?

Chapter 1

Getting Down to Business with VoIP

In This Chapter

- Getting over POTS
- Packetizing plain old telephone service
- Beginning with IPT on the LAN
- Reaching out with VoIP
- Uncovering the TCP/IP model

Technological innovation is hurling itself upon us once again. This time, it's coming in the form of improving the way we communicate, bringing with it new capabilities that change the meaning of the phrase *telephone call.* VoIP (often pronounced "voyp") is the name of this new communications technology.

VoIP, which stands for *voice over Internet protocol,* basically means voice transmitted over a digital network. Well, that isn't technically accurate because the Internet isn't strictly necessary for VoIP, although it was at first. What is necessary for VoIP technology is the use of the same protocols that the Internet uses. (A *protocol* is a set of rules used to allow orderly communication.) Thus, *voice over Internet protocol* means voice that travels by way of the same protocols used on the Internet.

VoIP is often referred to as *IP telephony* (IPT) because it uses Internet protocols to make enhanced voice communications possible. The Internet protocols are the basis of IP networking, which supports corporate, private, public, cable, and even wireless networks. VoIP unites an organization's many locations — including mobile workers — into a single converged communications network and provides a range of support services and features unequalled in the world of telephony.

Technically, IPT refers to telephone calls carried over the organization's local area network (LAN) such as a single building location, a campus-like network, or even a LAN within your home. When IPT crosses from the LAN to the WAN or any other external network, including other LANs operated by the same company at distant locations or the Internet, it becomes VoIP.

In the Beginning, There Was POTS

Before digital networking took off, everyone had to use the one and only *POTS,* which stands for *plain old telephone service* (honestly, it does). POTS runs over a network called the *PSTN,* or *public switched telephone network.* These POTS telephone systems use the tried-and-true method of telephone service known as *circuit-switched.* (See Chapter 2 for more about the history of POTS, the PSTN, and the operation of circuit-switched telephony.)

For customers, the costs related to the regulated circuit-switched PSTN remain much higher than they need to be. Consumers as well as companies that must rely on POTS on a daily basis know what the POTS way of telephony means to their bottom line. The good news is that VoIP is an alternative that can greatly reduce or eliminate POTS-related costs. (Chapter 3 fully details the recurring charges of the POTS way of doing telephony.) VoIP also enhances productivity, leaving more money in the budget to do other things besides pay telephone bills.

From POTS to Packets

VoIP technology enables traditional telephony services to operate over computer networks using packet-switched protocols. *Packet-switched VoIP* puts voice signals into packets, similar to an electronic envelope. Along with the voice signals, the *VoIP packet* includes both the caller's and the receiver's network addresses. VoIP packets can traverse any VoIP-compatible network. Because VoIP uses packets, much more information can be carried over the network to support and enhance your communication needs when compared to traditional telephony methods.

In a circuit-switched network such as POTS, routing is less dynamic than with a packet-switched network. In the POTS world, if a line is down, the call can't go through. In a packet-switched network, multiple routes can be established, and packets can travel any of the available routes. If one of the lines supporting the network is down, the packet can switch to another working route to keep the call up.

With VoIP, voice signals can travel the same packet-switched network infrastructure that companies already use for their computer data. Chapter 7 goes into more detail about dedicated packet-switched networks that support VoIP.

Eye for IP Telephony

VoIP also makes possible other services that older telephony systems can't provide. VoIP telephony services are *interoperable,* meaning that they work well over all kinds of networks. They are also highly *portable,* which means they will work with any IP-enabled device such as an IP telephone, a computer, or even a personal digital assistant (PDA).

IP telephony works by taking traditional voice signals and converting them to a form that can be easily transmitted over a local area network. Thus, the heart of IP telephony is the same as traditional data networking with computers. IP-enabled phones handle the voice-to-data conversion well, but don't be misled — implementing VoIP doesn't mean that everyone has to use IP-enabled phones. The best VoIP providers implement IP telephony in a manner that protects your investment in existing telephone equipment, even if you have analog telephone stations. (You'll find more on this topic in Chapter 10.)

All IP phones have one important thing in common: a built-in network interface card (NIC), just like a computer uses. The NIC is critical for any network device because it provides the device with a physical address and a way to communicate over the network.

The physical address supplied by a NIC is called a *MAC address.* MAC stands for *media access control.* The MAC address uses a standardized address and is usually represented by six hexadecimal numbers separated by dashes. For example, the following is a valid MAC address: 00-0A-E4-02-7B-99.

To support IP telephony, a server is typically dedicated to run the software used to manage calls. Servers are just like personal computers, except they have more memory, speed, and capacity. The server stores the database that contains all the MAC addresses corresponding to all the IP telephone extensions assigned to users. Depending on the size of the LAN and the number of users, you may use more than one server. For example, some LANs running IP telephony dedicate a server just for handling voice mail.

Depending on the size of the LAN, one or more devices known as switches are installed. These *switches* are boxes that have a series of ports into which all LAN-addressable devices ultimately connect. (Examples of LAN-addressable devices include computers, printers, wireless access devices, gateways, and storage devices.) Usually the switches are set up in the communications closets around the LAN, and they operate 24/7. All the switches are interconnected, often with fiber-optic cable.

In a nutshell, all network devices, including your IP telephone, must physically connect to the LAN through a port on a switch.

Calling over a computer network

Voice over Internet protocol is often taken to mean basically what it states: Voice traveling over the Internet. When VoIP was developed, it worked only with the Internet. Today, VoIP works on all other major network types, including those used throughout the corporate sector.

Making internal calls

When you want to call a coworker at your same location, you dial the phone number corresponding to the person's name. The signals are packetized and sent to the managing server, where the packet picks up the MAC address of the person you're calling. Next, the packet is forwarded to the switch, then to a particular port on that switch, and finally to the IP telephone connected to the port. The coworker's telephone rings. When the coworker picks up the receiver or answers the call, a virtual connection is established between the coworker and yourself for the life of the call. IP telephony does all this at lightning speed.

Making external calls

The process of calling a coworker at an offsite location varies only a little. The call is still initiated in the same way. But because the coworker is connected to a different LAN, the local server sends the call not to a switch located on your LAN but through the company's WAN (wide area network). This is where IP telephony technically becomes VoIP.

Each LAN in a multilocation network is connected to the larger WAN. If you're located at the company's headquarters in Pittsburgh, and you call a coworker located at the office in Los Angeles, your call begins as an IP telephony call on your LAN. It then travels from your LAN through a gateway, switch, or router that is programmed to re-packetize your call and encode the VoIP packet with additional information, such as the address for the destination LAN.

Network gurus refer to the process of packetizing your voice telephone call as *encapsulation*. A good analogy for this fancy techno-term is putting a letter into an envelope for mailing. The difference is that these encapsulated packets contain the content of the telephone conversation in digitized form.

To participate in the company's VoIP WAN, each LAN needs at least one edge device, such as a router, a switch, or a gateway. An *edge device* is just that — a device that sits on the boundary, or edge, of your local network and provides a connection to external networks. Depending on the company's network design, these edge devices can even have multiple interfaces that connect them to more than one outside network. The edge devices take care of all the IP telephony traffic going off-LAN by encapsulating the signals into packets, encoding the packets with the correct addressing information, and forwarding the packets out onto the WAN, where they make their way in a packet-switched manner to their respective destinations.

When the packets arrive at the destination LAN, the edge device on that LAN breaks down the VoIP packets and forwards them internally to the server that manages IP services. From this point, the rest of the process is similar to IP telephony services described in the preceding section: The phone rings, the person being called answers, and a virtual circuit is established between the caller and the receiver.

Gaining Flexibility with VoIP

VoIP is not just about making and receiving telephone calls; it's about a whole new way of communicating. Sure, it includes telephone calls, but there is so much more to the VoIP telephony picture. VoIP integrates most if not all other forms of communication. You can even run videoconferencing to your desktop.

With VoIP, your company enjoys increased productivity and customer satisfaction. These improvements are typically realized through the flexibility offered by enhanced calling features. A few calling features, such as voice mail and call transfer, have been around in the POTS world for quite some time. On the other hand, integrating data, voice, and video applications to run over a single network and work with wireless phones are more recent innovations made possible by IP telephony.

Following are some enhanced calling features made possible by IP telephony:

- **Vemail:** Before IP telephony and VoIP, you accessed voice mail through a telephone and accessed e-mail through a computer. With VoIP, you can read your voice mail on your computer screen and listen to your e-mail through an IP-enabled telephone. The new term for this converged feature is *vemail* (pronounced "v-e-mail").
- **Web surfing:** Because VoIP operates with the same set of IP rules and protocols that support Web-based applications, it is possible to access the Web with an IP-enabled telephone. If you have an IP telephone with a large enough screen, it can display Web pages or a list of your favorite Web links. For instance, you could use your phone to view your stock exchange trading status or the current weather forecast.

In an IP telephony world, these calling features (and many more) are available with no monthly recurring charges. VoIP, with all of its many benefits, is quickly replacing traditional POTS-based technologies. VoIP is even becoming a superior replacement for many former computer-only applications.

One of the big stories with VoIP is the many new and exciting features that increase your ability to be agile and mobile. You no longer have to say "I've got to get to a phone!" VoIP can be on your desk, computer, mobile phone, or PDA. It can be hardwired or have no wires at all. This flexibility is astounding to those familiar with traditional telephony.

If you have a mobile user base, be sure to check out IP soft phones. A *soft phone* is software that works on a laptop computer or pocket PC and provides most of the functionality of a traditional desk phone. If a user can connect to a network, the soft phone provides a way to reap the benefits of IP telephony regardless of location.

Looking at the TCP/IP Model

Many people marvel at the very thought that the POTS method of placing telephone calls can be replaced by a technology that essentially runs on the computer network. They are also startled by the many new and exciting features that come with VoIP. However, people also question how VoIP can possibly work and are a bit suspicious about whether VoIP can really live up to all the claims.

The answer can be found in the very same model that has been supporting data-only networking since the inception of the Internet more than twenty-five years ago: the TCP/IP model.

Pronounced "t, c, p, i, p," the model uses a five-layer approach to networking. TCP/IP is adapted to enable it to also support VoIP. TCP/IP has proven to be just as effective with packetizing telephony as it has been for many years with packetizing computer data.

To fully understand VoIP, it pays to know a little about the technical underpinnings that make it work over the network of your choice. In this section, I describe the layers of the TCP/IP model in relation to computer networks. Then I insert into this content the parts that change when TCP/IP supports VoIP.

TCP/IP layers

TCP/IP is first and foremost a group of networking protocols. *Protocols* are the rules that govern how network traffic gets packaged electronically for transmission over a network. Some TCP/IP protocols are used strictly for data networking, some are used strictly for VoIP telephony, and some are used by both data and VoIP. Each protocol corresponds to one of five possible layers that make up the TCP/IP model:

- **Application:** Special protocols at this layer ensure the quality and deliverability of VoIP packets.
- **Transport:** The user datagram protocol (UDP) at this layer transports the VoIP packets from start to finish, which in this case means from caller to receiver and vice versa.
- **Internetwork:** At this layer, IP addressing is added to the packet. Every VoIP phone or computer acting as a VoIP phone gets a unique IP address that routes delivery of VoIP packets to and from the caller and receiver during the life of the call.
- **Network interface:** At this layer, MAC addressing is added to the packet. (The MAC address is supplied by the NIC required for all network devices.)
- **Physical:** This layer converts all packets to electro or electro-optical signals to be carried over the local or external network.

Each layer is associated with one or more protocols. A packet must traverse all five layers: once when the packet is sent and again when it is received.

Basically, the VoIP packet originates with the caller. The packet travels down all five layers on the caller's side of the network and gets packaged with the correct protocols at each layer. After the packet reaches the lowest layer, the physical layer, it is sent over the network to its destination. When the packet reaches its destination, it makes its way up through the layers and gets unpackaged. When it reaches the application layer of the receiver, the packet is translated into a voice signal that the receiver hears.

TCP/IP differences

TCP/IP protocols are applied a little differently depending on whether you have a traditional data packet or a VoIP packet. Figure 1-1 illustrates the packet breakdown and corresponding layers involved in a TCP/IP network connection for a standard Web application, which uses a traditional data packet.

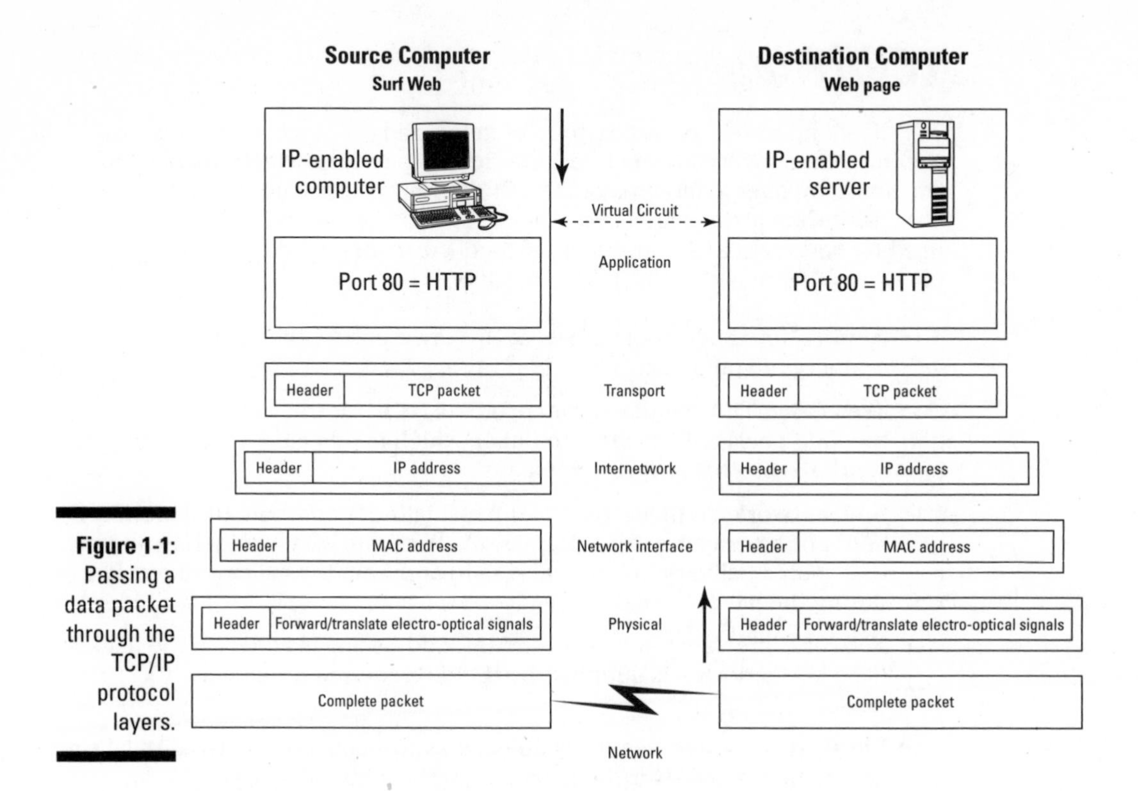

Figure 1-1: Passing a data packet through the TCP/IP protocol layers.

Note that the transport layer of the packet uses the familiar TCP protocol to construct the packet exchanges between the source computer and the destination computer (in this case, the Web server).

Figure 1-2 illustrates the layered protocol stack shown in Figure 1-1, but this time applied to a VoIP call. Note the *header* at each layer's version of the packet (except the application layer). The various headers identify what layer the packet is on during its travel from caller to receiver.

If you compare Figures 1-1 and 1-2, you notice two protocol differences, at the application and transport layers of the TCP/IP model. Other than these differences, a great deal of symmetry exists between voice and data using TCP/IP. That's why VoIP can behave like a telephone while delivering many computer-related functions.

Application layer differences

The first difference between the VoIP implementation of TCP/IP and the traditional data implementation is in the application layer. In a VoIP call, the application layer utilizes the following three protocols:

- **NTP:** Network time protocol. This protocol enables timing, which helps ensure that the signals are transmitted and received within the proper timeframe to assure quality.
- **RTP:** Real-time transport protocol. This protocol provides end-to-end network transport functions for digital voice signals encapsulated in the VoIP packet.
- **RTCP:** Real-time transport control protocol. This protocol monitors voice signal delivery and provides minimal control functions to ensure the delivery of packets.

All three of the application layer protocols combine, at nanosecond speeds, to deliver VoIP voice packets.

Transport layer differences

The second difference between the traditional data implementation of TCP/IP and the VoIP implementation is in the transport layer. The lion's share of computer data networking uses the TCP protocol at the transport layer. For VoIP, the transport layer uses UDP, user datagram protocol. (UDP is used also for real-time videoconferencing networks.)

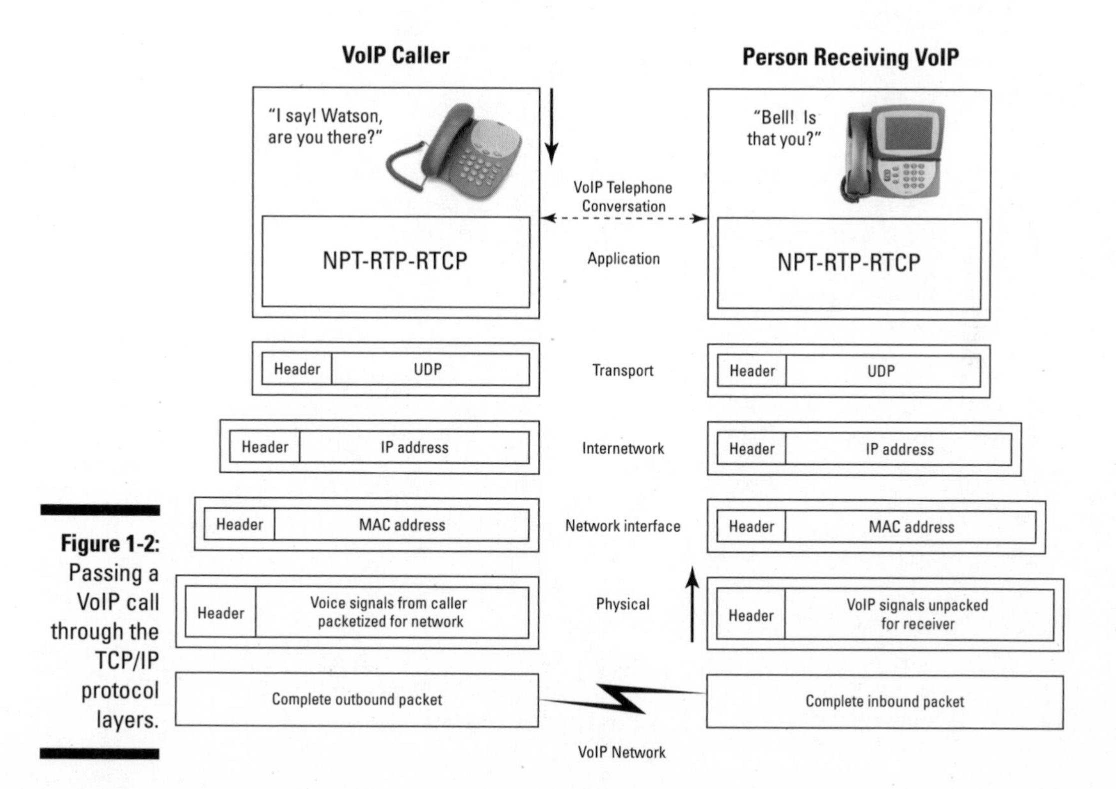

Figure 1-2: Passing a VoIP call through the TCP/IP protocol layers.

TCP is slower than UDP, but it provides guaranteed delivery of its computer data packets. Keep in mind that we are measuring speed here in nanoseconds. Even if it takes a long time for the packets to reach their destination computer, eventually TCP ensures delivery.

Because voice is a real-time application, it is more important that the voice packets get to the receiver as quickly as possible. That is why UDP is by far the hands-down favorite to provide the transport layer for VoIP networks.

Chapter 2

VoIP: Not Your Father's Telephone Service

In This Chapter

- Uncovering the origins of VoIP
- Telephoning using analog lines
- Going digital
- Networking analog and digital
- Digitizing the public telephone network
- Raging war between circuits and packets
- Replacing POTS phone systems with VoIP
- Discovering how private telephone systems reduce POTS costs
- Watching network types multiply and morph

Voice over IP represents a significant change from the traditional way that telephone calls have been handled until recently. Even so, the genesis of VoIP is rooted in the history of networks, specifically, the history of the circuit-switched phone network.

This chapter explores just a bit of that history. It offers a whirlwind tour of how phone systems got to where they are today and how that relates to VoIP. By understanding the way that phone networks relate to things such as regulated phone costs, you'll find it easier to grasp the huge cost savings that can be realized by converting to VoIP.

Mr. Bell

Believe it or not, the roots of VoIP go all the way back to the 1870s. In 1879, Alexander Graham Bell forgot his Internet password and, knowing that his assistant had stashed it away, uttered the famous words "Watson! Are you there?" He never got on the Internet, but he did prove that the human voice could be carried electronically over a pair of wires. He also demonstrated that the endpoints for these wires had to be connected to the right equipment — hardware that he invented. Mr. Bell's inventions ushered in an age of communication that made the world much smaller than it had ever been before.

When Mr. Bell invented the telephone and thereby gave birth to the telephone network, VoIP was not even a consideration. (Truth be told, the idea of a network wasn't yet a consideration either.) Other inventions would be required before VoIP could become a reality.

The first telephone equipment was analog. Historians and technicians alike have labeled the first phone service *POTS*, or *plain old telephone service.* VoIP won't function very well over a POTS system; it requires a digital network.

Digital networking for telephones was invented in the 1920s, but the first digital networks would not leave the laboratory until much later, in 1964. Today, most phone companies in the United States have updated their equipment to include digital service.

Over time, the POTS network gave way to the *PSTN,* or *public switched telephone network.* (The terms PSTN, public switched telephone network, public telephone network, and phone network are used synonymously.)

Although it occurred in what seems like the ancient past, Alexander Graham Bell's work is important in understanding VoIP. The POTS network that began with his invention has grown into the largest circuit-switched network in the world. It also has become an expensive network, with individuals and companies spending hundreds of billions of dollars each year for communication services.

VoIP, which was developed in 1995, is gradually replacing the PSTN. Some view the PSTN as the antithesis of VoIP, but it still remains the standard of quality by which VoIP is measured. For instance, people often ask whether VoIP provides voice quality as good as what is delivered through the PSTN. Most of the factors used to evaluate the quality of VoIP are based in some way on the PSTN, so understanding a bit about the older networks is important.

Analog Telephone Circuits

As mentioned, phone technology originally was analog, from start to finish. *Analog modulation* is the technique used to convert sounds (such as your voice) into an electromagnetic form. The analog circuitry of the POTS telephone transmitter converts the voice patterns coming from the caller's mouth into continuous electromagnetic signal patterns. These patterns are carried on a telephone line circuit, sometimes called a *trunk line,* where they are carried to the terminating end of the circuit. There, analog circuitry converts the signal back into audible sounds so they can be understood by humans.

A good basic illustration of a POTS circuit can be found in an old elementary school science experiment. My fifth-grade science teacher, Ms. Davis, had us punch a hole into the end of two tin cans and connect them using a long string. If we held the string taut, Jodie Schnickmeister could whisper into one can and I could hear her in the other. (I used to love it when Jodie whispered in my ear.)

This simplistic experiment taught the basics of the POTS network: Sound was converted to an analog signal (vibrations) that was carried over the taut string to the receiving can. The string, in turn, vibrated the can and converted the analog signal back into audible sounds.

Telephony Goes Digital

Scientists, never content with two tin cans and a string, looked for different ways to transmit sounds over long distances. The pioneering work of Harry Nyquist in the 1920s gave us the basics of sampling theorem. In the 1940s, Claude Shannon would mathematically prove Nyquist's sampling theorem. Their work is the foundation for what we now call digital networking. Basically, they proved that you could take the analog signals of any POTS call and convert them to digital form. This meant that POTS calls could originate in analog form, be converted to digital form, and be transmitted on the PSTN using the now familiar ones and zeroes of computers. Digital networking had arrived, setting the stage for the beginning of VoIP.

The work of Nyquist and Shannon led to many telephone and computer network inventions. For example, Nyquist is credited with the patent that led to the first coder-decoder, or *codec,* device. Codecs can come in many sizes and shapes and are often found in the electronic circuitry of large networking devices. Codecs basically convert analog signals to digital form and vice-versa. Nyquist's work led to the design of many other networking devices such as dial-up modems, high-speed broadband modems, IP routers, and VoIP gateway servers.

The ability to convert analog signals to digital form also led to the development of several types of computer networks. From the early 1960s to the present day, several types of digital networks, including fiber-optic-based networks and wireless networks, have emerged in support of computers and telephone systems. Today's digital networks, regardless of the form they take, are capable of supporting VoIP telephony. We cover network types beginning in Chapter 4.

Combining Analog and Digital

When digital networks were introduced, the phone companies wanted to use them right away because they provided a more efficient means of transmitting signals all over the place. (Digital networks could carry data much faster than analog networks.) The phone companies were presented with a problem, however: how to make existing analog phones work with a digital network.

The answer was to use a codec to convert the analog signal to digital. But where should the conversion take place? At the phone company's facilities or at their customer's location? In the early years of the digital revolution, the conversion took place at the phone company's facilities, which allowed the phone company to utilize the existing wiring between their facilities and the customer's location. This wiring between a phone company facility and a customer is often called a *local loop*.

Over the years, the codec has been pushed closer and closer to the customer, all in an effort to make the phone network as close to 100 percent digital as possible. Most parts of the PSTN remain a combination of analog and digital. Customers pick up a phone, which converts audible sound into analog signals. These signals are carried over the local loop to the phone carrier's facilities, where they are converted to a digital signal. The signals are forwarded to the receiver's end. After the signals are received by the last piece of carrier equipment (closest to the end customer), they are converted back to analog form. The analog signals go into the receiver side of the POTS telephone and are heard as a replica of the caller's voice. Figure 2-1 illustrates how a phone call is transmitted over the PSTN.

Calling Pennsylvania 6-5000

In the 1940s, a consortium of leaders in the telecommunications industry and in government standardized how customers would be assigned telephone numbers. The telephone number identified a specific pair of wires out of millions of pairs of wires, and a specific phone company switch out of thousands of such devices. The term *circuit-switched* describes this setup of circuit wiring, switching devices, and telephone number assignment. The PSTN is sometimes referred to as the circuit-switched or switched network.

Because today's public phone system is still circuit-switched, it still relies on the same basic system for telephone number assignment. VoIP introduced dramatic changes in how the network is used and, over time, VoIP could force changes in how numbers are assigned. With VoIP, phone numbers are no longer tied to specific wires and switches. VoIP routes calls based on network addresses, and phone numbers are simply used because that is what people are familiar with. (VoIP takes care of translating a phone number into a network address.) In the future, as more and more people adopt VoIP-based systems, we may see dramatic changes in phone numbering.

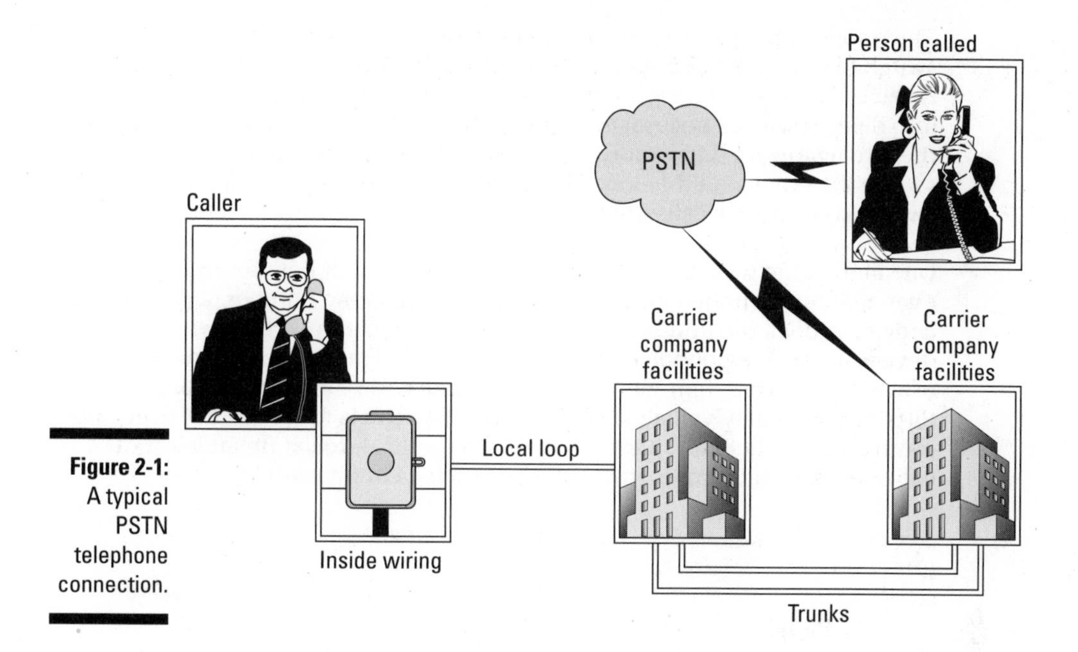

Figure 2-1: A typical PSTN telephone connection.

Today, a customer can have either an analog or a digital phone. Depending on customer location and end-user equipment, the POTS call can be 100 percent digital.

For more than a century, POTS remained the dominant form of two-way telecommunications. During that time, however, POTS-based telephone systems changed dramatically in the number, length, diameter, and type of wire or cables used and in the types of telephone equipment both at the customer end and at the carrier's facilities.

Digital Telephony Invades PSTN Territory

When digital networks were implemented back in the 1960s, the telephone carrier companies began using a technique that permitted them to accept analog telephone calls coming into their switching facilities and convert those signals into digital form for transmission on their shiny new networks. They had not yet made the leap into packetizing telephone calls, which is what we have today with VoIP. At the time, they thought it best to keep the circuit-switched telephone carrier network physically separate from the evolving packet-switched computer network.

The phone companies were able to make other improvements to circuit-switched telephone services. After their circuit-switched carrier network received the caller's telephony signals, they were able to convert the signals into digital form, as necessary. They discovered that digital signals allowed them to aggregate many more calls onto a given circuit and through a given switch than they could before. This enabled them to streamline how circuit-switched telephone calls could be made.

One innovation was the addition of area codes, which help to process calls over a circuit-switched network. The entire telephone number, including area code, identifies the number of circuits and the location of the switching devices for a given phone on the PSTN network. For example, consider a call originating in Mountain View, California (area code 415) to a person in Pittsburgh, Pennsylvania (area code 412). The call is switched out over three physically distinct circuit switches — 415 to 412 — to set up and carry the call. Figure 2-2 illustrates the routing of such a circuit-switched call.

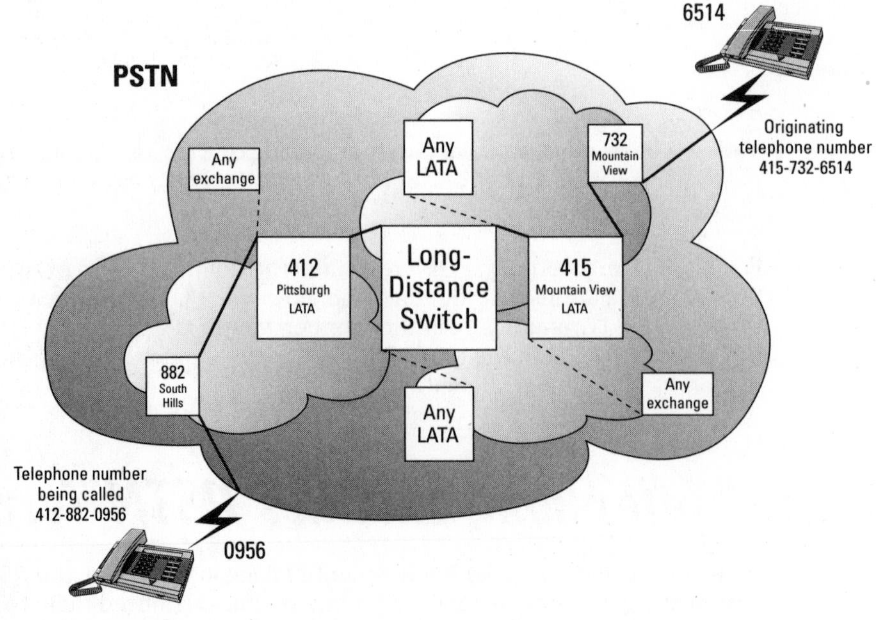

Figure 2-2: Routing a call from area code 415 to area code 412.

If these two locations are on the same computer network, and both are using VoIP, the call could be carried over the company's computer network in packet form. This process is known as *on-net* VoIP telephony. None of the packets would touch the PSTN. There would be no toll, regulatory, or metered charges for this long-distance telephone call. Figure 2-3 illustrates the routing of such a packet-switched VoIP call.

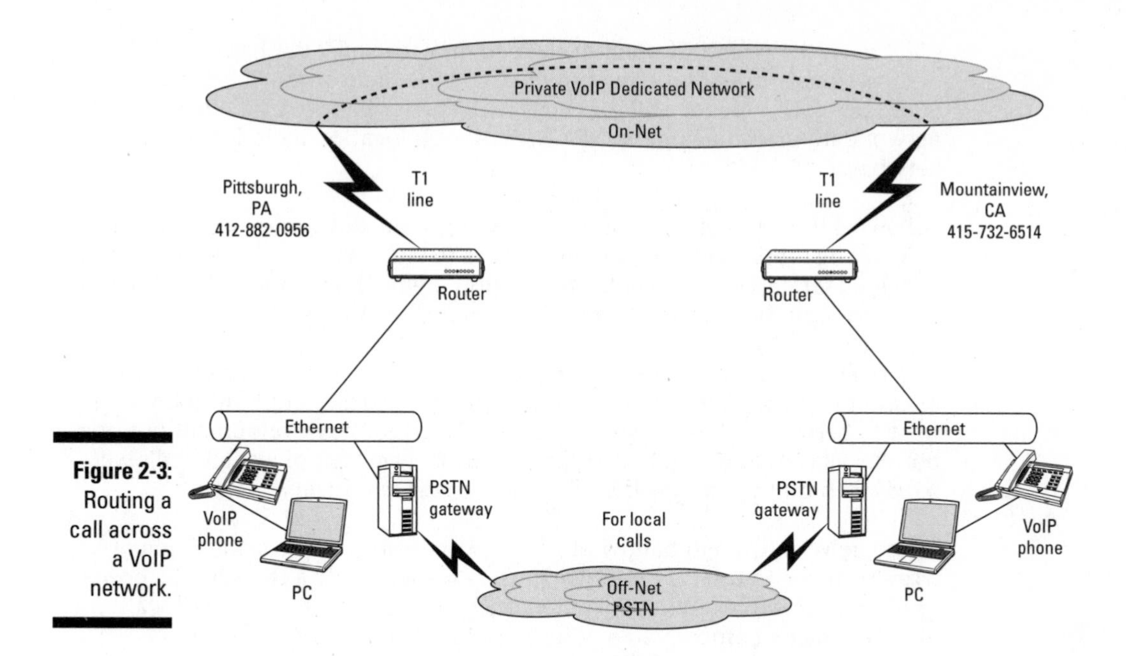

Figure 2-3: Routing a call across a VoIP network.

The circuit-switched network gets organized

As circuit-switched networks continued to evolve, other technologies were developed that helped the carriers manage their telephony operations. Carriers began offering more types of POTS access and POTS carrier services.

The early forms of local and long-distance carrier services had to be redefined according to where the carrier company had facilities to terminate the circuits and transport lines, as well as where they might install their facilities. In addition, government regulation of telecommunications picked up. The concept of a *local access and transport area,* or *LATA,* as a geographical designation was defined. Eventually, the entire map of the United States would be developed into thousands of LATAs. You can usually identify a particular LATA by the area code associated with a telephone number.

The big advance with LATA was that it helped carriers get organized in a manner that would let them offer other types of carrier services, including those outside the circuit-switched services of the PSTN. For example, a numbering plan was developed that identified any circuit or access transport by its area code and the prefix of the main telephone number. The area code became known as the NPA, for numbering plan area, and the prefix became known as the NXX, for number exchange. For example, the NPA-NXX 412-882 is the area code and prefix for the Pittsburgh 412 LATA and switch 882, located in the South Hills of the Pittsburgh 412 LATA.

What does the NPA-NXX number have to do with VoIP? In Chapter 7, I discuss dedicated networks that have proven to be the highest quality of service (QoS) network type for VoIP networks. All carrier lines for a dedicated VoIP network are priced using the NPA-NXX of each location included in the network.

Following the development of digital services, the corporate sector began demanding more bandwidth from carriers to support their networks. It didn't take long for carrier companies to develop digital, high-bandwidth transport lines that could meet the diverse needs of the corporate sector.

These newer transport lines would be digital all the way from a customer's location A to the same customer's location B, regardless of how many miles were in between. These customer demands led to the development of transport services that multiplied exponentially the amount of available digital bandwidth that could be offered to the corporate sector.

These newer ultrahigh-bandwidth transports were not the same kinds of wires as those in the POTS network. They were usually a thicker-gauge wire or fiber-optic cables. When installed, they connected two or more locations of a customer's company in a point-to-point fashion versus the circuit-switched method of the PSTN.

These developments contributed to the emergence of private dedicated networks, which in turn ensured that VoIP would be here to stay. (As you find out in later chapters, VoIP becomes a viable option for companies only when used over dedicated networks.) Eventually, on-net VoIP over dedicated networks will replace expensive circuit-switched calling over the PSTN.

The digital services carrier network

The new types of digital lines installed by the carriers began to form a new physical carrier services network. The lines did not cross-connect or intersect with any of the millions of circuit-switched lines that are in place and continue to be installed by the carriers. At the carrier company's facilities, newer types of fully digital equipment terminated these digital lines.

This new carrier services network was called the *digital services carrier network.* (It is also known as the *digital signal carrier network,* or simply as the *DS.*) This network used higher-bandwidth digital lines and operated with packet-switched protocols to network computer data. (For more on protocols, which are simply rules for using the network, see Chapter 1.)

Soon thereafter, the DS network was defined based on its fundamental unit of bandwidth, known as the *channel.* The smallest channel unit provided a bandwidth of 64 Kbps (64 thousand bits per second). This channel was called a DS0, pronounced "D–S–zero." (Many computer gurus start counting with 0; it's a binary thing.) DS0 became the base unit of bandwidth from which other dedicated transports were defined. (In Chapter 7, I cover the popular DS standards in detail.)

The DS network also introduced to the telecommunications vocabulary another term that characterizes most of the transport mechanisms that are not part of the older circuit-switched network. Because the lines used by the DS carriers were installed between private customer locations and the public at large could not use or connect to them, DS lines became known as *dedicated* to the customer leasing them. The entire series of DS standards eventually became known as the *dedicated carrier services network.*

War Breaks Out Between Circuits and Packets

The corporate sector's thirst for leasing dedicated DS lines was unquenchable. Soon a dilemma emerged as to how to distinguish the circuit-switched network and the newer dedicated network, which used packet-switching technology. Since its inception, the circuit-switched network was a public carrier services network. The DS network was being called dedicated, or private, because no one but the customer paying for the DS lines was permitted to use them.

Public versus private a.k.a.

Confused over the different terms for the PSTN and dedicated networks? The following list comes to the rescue, showing the most popular also-known-as terms for both.

PSTN	DS
public	private
public switched telephone network	private dedicated transport network
public telephone network	digital services network
switched network	dedicated network
circuit switched	packet switched

It wasn't a great leap to make the distinction between public and private types of services. At this point, the name *public switched telephone network* (PSTN) began to be used to characterize the circuit-switched network. Eventually, the PSTN would be referred to by the telecommunications industry as simply the *switched network*. The DS name stuck with the network that provided private dedicated transport services. Eventually, the DS network would be referred to as simply the *dedicated network*.

Figure 2-4 illustrates the physically separate PSTN and DS networks.

POTS telephony continues to use circuit-switched protocols that don't packetize telephony signals. (See Chapter 1 for an explanation of packets.) POTS signals travel from one line to the next line on a given circuit of lines, just like in the fifth-grade science experiment using tin cans and a string. Another good way to understand circuit-switched protocols is to think about a railroad system. Trains must switch tracks along a circuit of tracks based on the destination of the railroad cars traveling over the tracks. The direction of the train is determined by the physical tracks that the train uses. Figure 2-5 illustrates such a circuit-switched train.

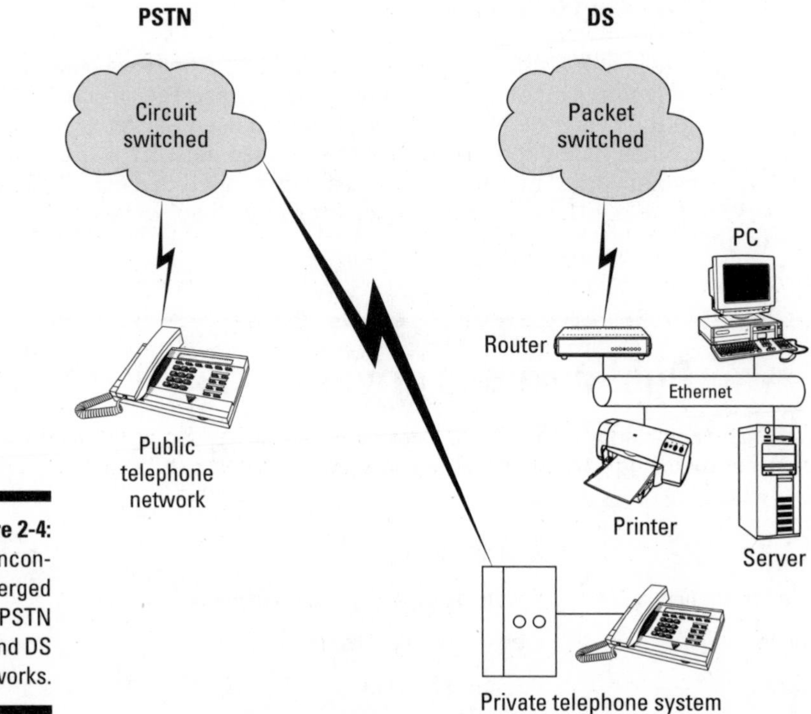

Figure 2-4: Nonconverged PSTN and DS networks.

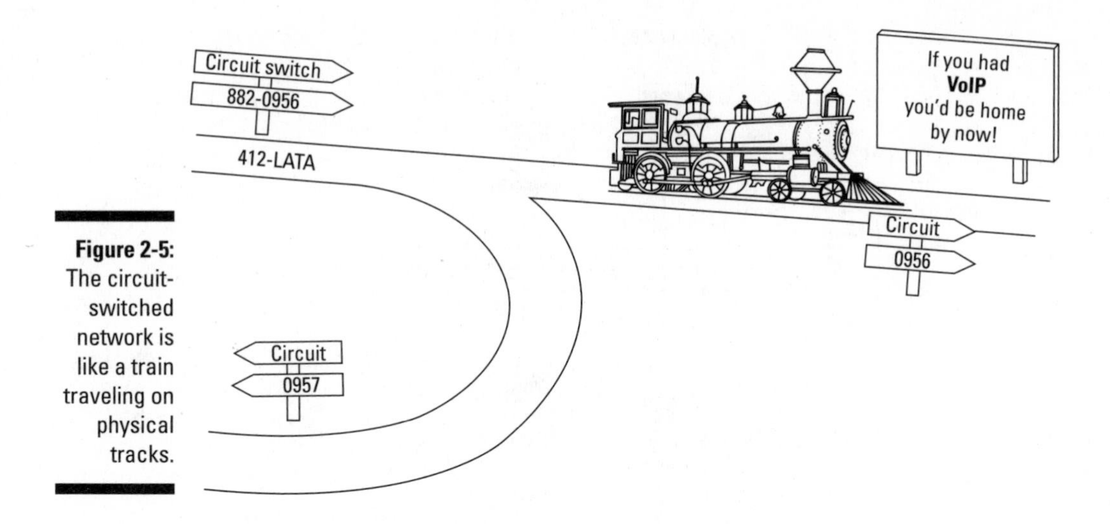

Figure 2-5: The circuit-switched network is like a train traveling on physical tracks.

VoIP technology has enabled telephony signals to run over dedicated networks using packet-switched protocols. One of the preferred methods of running VoIP in the corporate sector is to use dedicated lines. Instead of being primarily dependent on the PSTN for its telephone service requirements, companies using VoIP protocols can send and receive telephone calls over their private computer networks. Using VoIP, voice signals can be packetized in a manner similar to computer data packets.

VoIP includes the caller and receiver's network addressing information in the packets sent over the network. If a given circuit on the network is down, VoIP packets can switch to another computer network circuit because the packet is not dependent on the circuit itself for directions. In the previous example, the circuit-switched train is switched solely by the tracks it travels. If the train runs into a broken track, it can't continue to travel to its destination. VoIP packets can have many alternative routes because the destination address inside the packet tells the network where to route the packet.

Most companies today use packet-switched networks for their computers and separate circuit-switched networks for their voice calls. Figure 2-6 depicts this typical scenario.

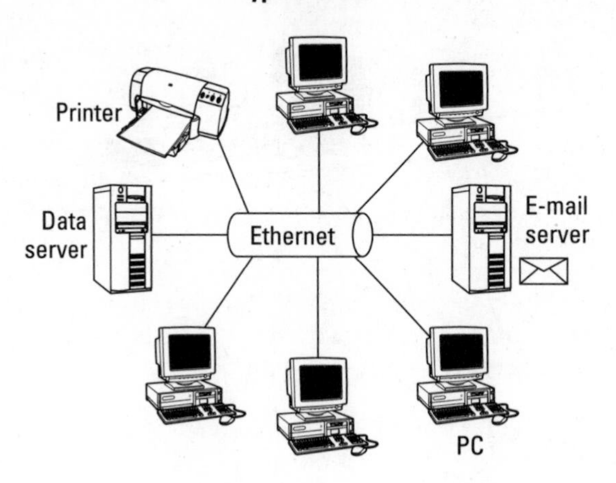

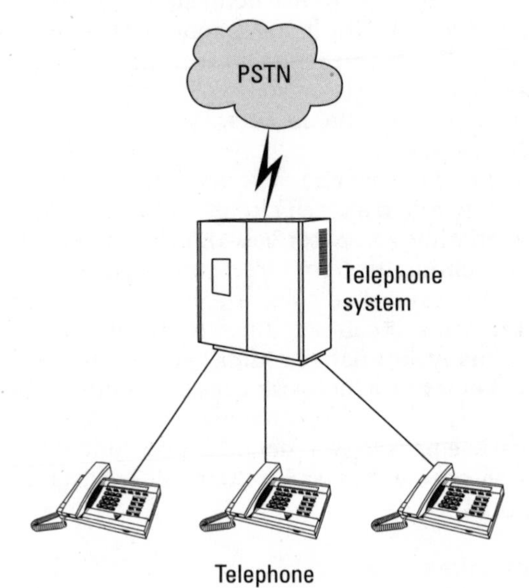

Figure 2-6: Companies typically use noninte-grated networks for data and voice.

Private Telephone Systems Reduce POTS Line Costs

Computer data networks and circuit-switched voice networks are completely separate, with individual staffing, billing, maintenance, and accounting systems. Although the maintenance costs of computer networks are affordable for most companies, the recurring charges for traditional forms of telephony are huge for small, medium, and large multilocation companies. VoIP is designed to converge (integrate) a company's voice needs onto the company's existing computer network. If a company does this, they can eliminate most (if not all) recurring circuit-switched telephony charges.

In the past, the POTS world had only two types of services: local and long distance. Local service covered the entire metropolitan area, with no distinctions for the various levels of toll service that we have today. In the early days of the telephone, long-distance cost customers dearly. A call from New York to the west coast might have cost $3 to $4 per minute. Today, that same call might cost a consumer $.02 to $.05 per minute and a corporate caller $.01 to $.03 per minute. The corporate customer is most likely on some sort of dedicated private network consisting of a phone system connected to the PSTN.

It might appear that the cost of telephony today is dirt cheap in historical terms. This would be a mistaken conclusion. In addition to the carriers getting more organized and the government increasing its regulation of the telecommunications industry, many changes have evolved. These changes have increased your bottom-line telephone bill and increased the number of line items on that bill.

Now, instead of just two types of phone service offered on the PSTN (local and long distance), we have five: local, intralata, intrastate, interstate, and international. Each of these is discussed in detail in Chapter 3. These five services are based on the origin and destination of a call, using the LATA and NPA-NXX to determine those locations. In addition, the same system is used by the government to place various surcharges and fees on each telephone access line.

No one would argue that the quality of carrier-switched telephony is excellent. However, the system that has evolved for charging telephony customers leaves much unsaid and a lot to be desired. Except for local calling, VoIP can reduce or eliminate the charges of the other four categories.

To lessen the burden of newer and diverse telephone costs, many companies have acquired their own POTS-based telephone systems. Company-sponsored telephone systems can reduce the monthly bill that consumers and companies pay for telephony services. Four different telephony system models have evolved in the past three decades.

The first model, POTS, has already been described; it is the use of telephony access lines and carrier services over the PSTN through a carrier. The other models are the Centrex, KTS, and PBX models. Each of these are discussed in this sections.

The Centrex model

The second model is the central office exchange service, or Centrex, model. Centrex is physically set up the same as the POTS access line model. Like POTS, Centrex uses the same physical twisted-pair copper lines.

The difference between the POTS and Centrex models lies in how the line is terminated at the carrier company's facility. Instead of getting switched into the PSTN directly, the Centrex line first goes to the more intelligent mainframe-level telephone system owned and operated by the carrier. From there, the system can provide the customer with many more features not directly available on a plain POTS line. To get these features on a POTS line, the customer typically has to pay for each feature. Centrex provides a bundle of features with little or no added charges.

Because you get more with Centrex, you pay a little more for the line on a per-line per-month basis. Centrex is a good alternative for companies operating out of temporary facilities (such as a lease situation) and for companies that can't or don't want to maintain a full-blown telecommunications infrastructure. When you lease a Centrex system, maintenance is usually included, which reduces the need for skilled staff on the company payroll. Figure 2-7 shows how the addition of a Centrex system modifies the model originally shown in Figure 2-1.

The good news is that a VoIP solution exists for the more traditional Centrex situation: VoIP Centrex. You need a computer network in the temporary premises, but that is an expense you can unplug and take with you when you move to your permanent location. With VoIP Centrex, you can start your VoIP network, acquire features galore, and get the maintenance services you need. When you move to your new, permanent location, you simply plug in your network, and you are up and running.

Figure 2-8 shows the addition of VoIP Centrex to the mix.

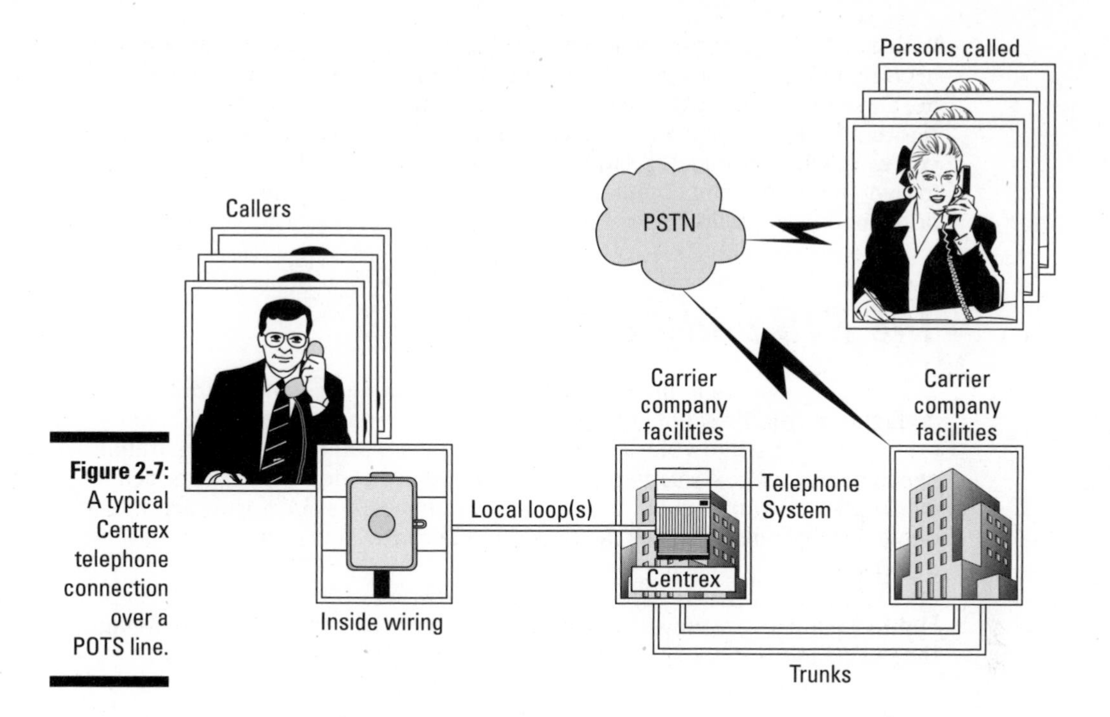

Figure 2-7: A typical Centrex telephone connection over a POTS line.

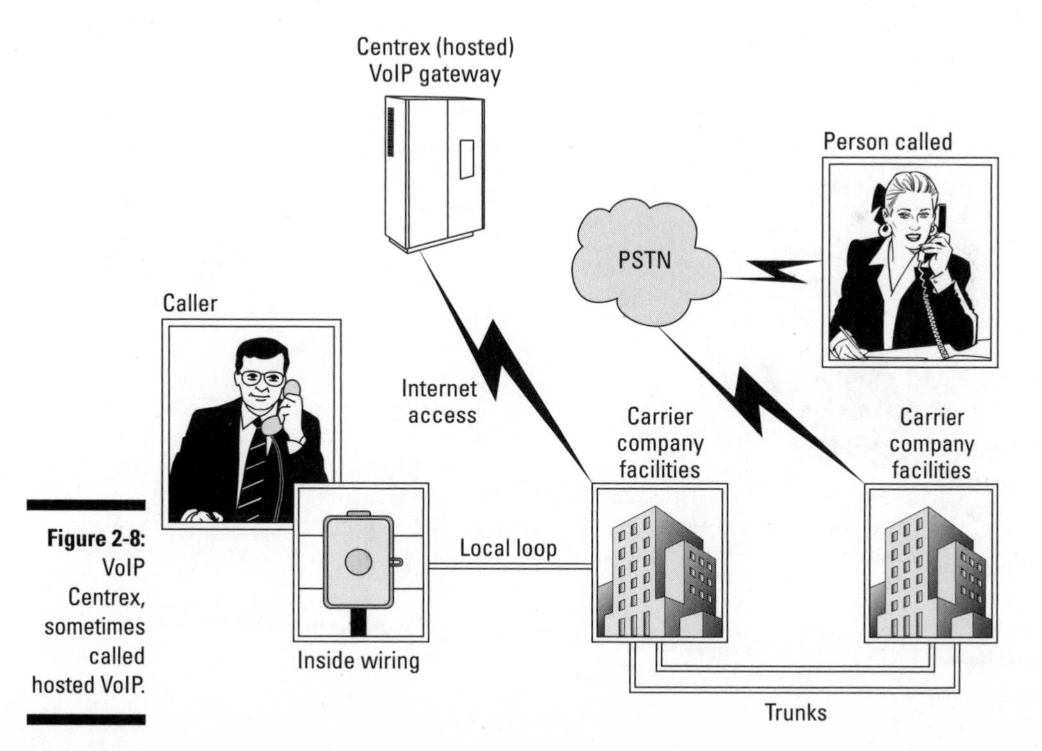

Figure 2-8: VoIP Centrex, sometimes called hosted VoIP.

With any Centrex option, you pay more per month in return for avoiding the need to sink costs into your own infrastructure. Also, by using the Centrex host's facilities, you get a rich feature set with no additional monthly charges. Another benefit is that you can walk away from a Centrex solution anytime you want without penalty. Although some smaller companies keep Centrex forever, most growing companies eventually convert to one of the models described in the following section.

The KTS and PBX models

The other two system models are private telephone systems installed on the company's premises. Low-volume customers often use a key telephone system, or KTS. High-volume, larger companies often use a private branch exchange, or PBX. These two are a departure from the POTS-line model, where a line is run to each phone on the premises. As such, they are also a departure from the Centrex model, which uses the same type of access line as POTS.

Figure 2-9 illustrates how a typical private telephone system would change the phone mix.

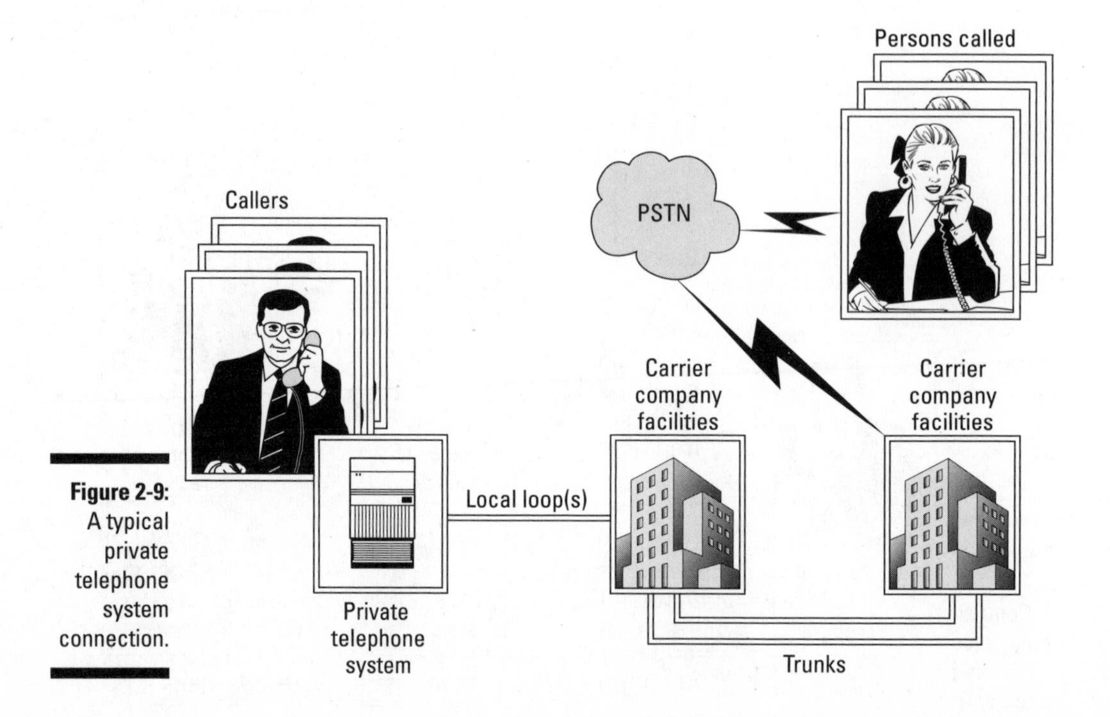

Figure 2-9: A typical private telephone system connection.

One big benefit of private telephone systems is that they reduce the number of required access lines. For example, the industry standard is one access line for every six to eight employees who have a telephone. Any reduction in the number of access lines represents an enormous cost benefit for companies when compared to the POTS or Centrex models. In addition, private telephone systems enable the company to provide most traditional call features, such as voice mail, call forwarding, call transfer, and conferencing, to any telephone in the company — at no added cost.

Private Systems versus VoIP

A private telephone system approach can't begin to compare to a VoIP model in terms of savings. Your guide should be "How much telephone calling traffic, across all five regulated PSTN charging categories, do you or your company have each month?" If your monthly call volume, which is charged by the minute for each line across each charging category, is substantial, a private telephone system model reduces your recurring charges because you use fewer lines. However, VoIP can reduce your recurring charges even further, as you'll discover in the next chapter.

Following is a list of cost benefits and features that your company can gain by converting to its own telephone system.

- Greatly reduced number of access lines
- Reduced recurring carrier charges
- Reduced access line fees and surcharges
- Reduced access line taxes
- Elimination of call feature charges
- Greater managerial control of telephony systems and services

There is no doubt that moving to a private telephone system saves a company significant money when compared to a POTS access-line model. But keep in mind that all of these cost benefits are based on reductions in the number of lines required or lower costs for features priced on a per-line basis. These benefits are also realized with any VoIP model.

The conventional telephony models described in this section, KTS and PBX, don't remove the problems associated with telephone costs. They only minimize them by adjusting the number of access lines or calling features you need to pay for. A VoIP system, on the other hand, represents a fundamental change in telephony, and thereby offers huge cost savings, feature enhancements, and productivity improvements. VoIP eliminates the need for most access lines. (A few POTS lines are always required in any building.) VoIP eliminates also the noncarrier costs (that is, your maintenance costs), line fees, and government surcharges that come with those lines. And VoIP runs on the computer network, which is usually already set up.

Converging Networks

In Figure 2-10, you see a packet-switched computer network running VoIP and connected to the traditional PSTN. Note the absence of any POTS lines or private telephone systems (KTS or PBX) under the DS carrier service network cloud. All telephone calls are originating on the company's computer network using VoIP. Only calls destined for the PSTN are diverted off the company's network. These types of calls, referred to as *off-net,* are the only calls that may be associated with a recurring service charge. For customers and companies running VoIP, off-net calls typically are only the calls that go to the public network (for example, to order a pizza or to call 911). On-net calls require no additional lines over the existing computer network setup and, unlike POTS-PSTN calling, have no additional recurring charges.

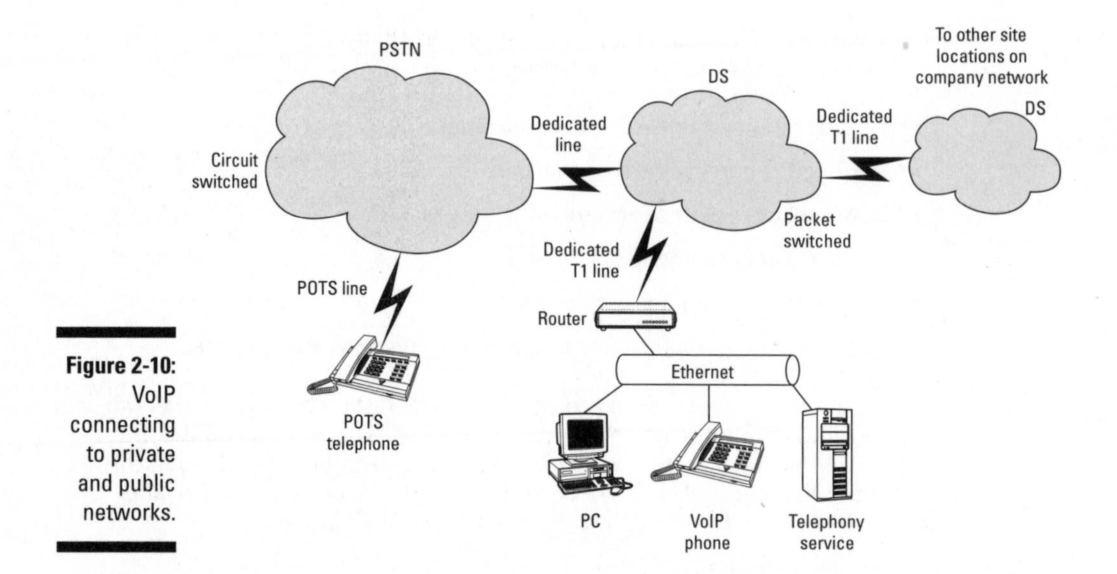

Figure 2-10: VoIP connecting to private and public networks.

A smaller company with just a single location may use VoIP to connect their computer network to the PSTN to support calls that must travel off their private network. This solution requires installing a VoIP gateway that can convert the company's on-net traffic to circuit-switched telephone calls that have an off-net destination. For example, a call from your desktop phone to the grade school that your daughter attends is likely to be a local PSTN call. Figure 2-11 shows how a VoIP network works with a gateway.

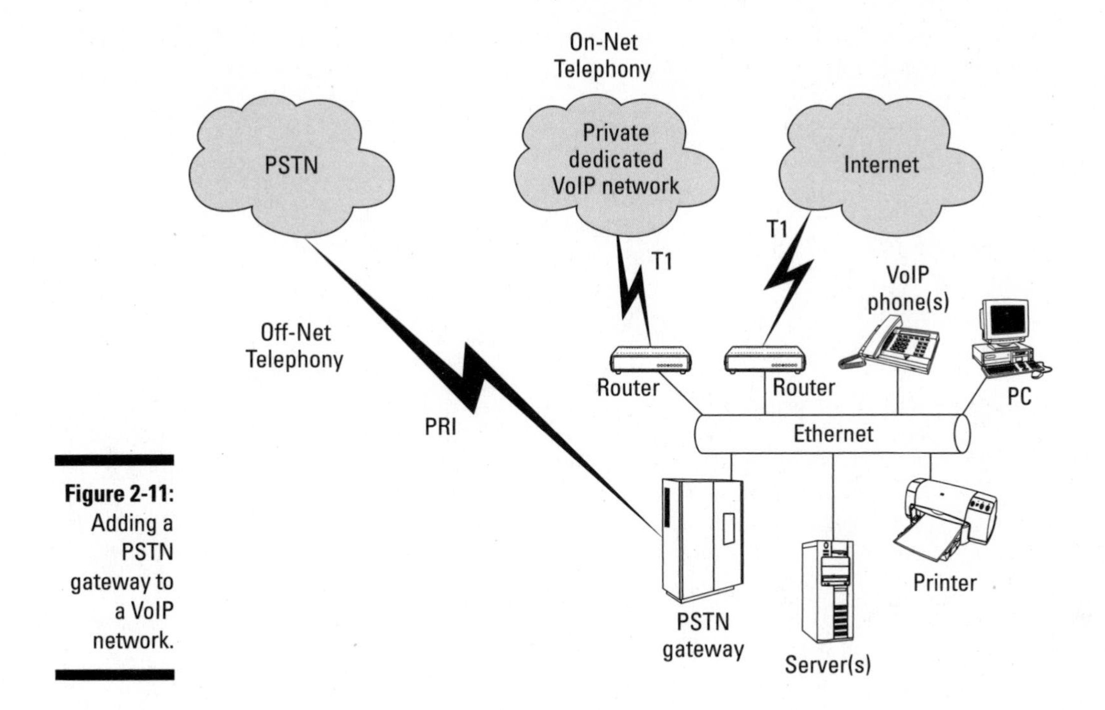

Figure 2-11: Adding a PSTN gateway to a VoIP network.

Converting VoIP telephony traffic to run on the PSTN is certainly a big benefit of VoIP, but many more exciting features and benefits are explained in the chapters that follow. For example, in Chapter 3, I describe how VoIP reduces or eliminates those nasty, pay-by-the-minute service charges and other recurring charges such as regulatory fees and taxes.

Chapter 3

Everything You Need to Know About Charges

In This Chapter

- Getting access to the network
- Understanding the costs of service categories
- Surviving your monthly bills
- Analyzing savings with a case study

In the old days of making calls with a telephone (remember, just last year), you paid for a phone line. Your company may have had one network to handle dozens or even hundreds of phone lines coming into your business and another network to handle computers. Now companies can converge both networks into one. By using VoIP over a private data network, your company can bypass the older, more expensive way of using the public circuit-switched network.

Although some local telephone lines may be necessary, you can reduce or eliminate the cost of your older telephony infrastructure, the total volume of call minutes per month, line-related regulatory fees and taxes, and therefore most if not all of your total phone bill. When was the last time your phone bill was *less* than you thought it would be?

This chapter describes the bottom-line savings of using VoIP. You see how traditional calling methods hit you with charges for everything under the sun and how calling with VoIP can change this for the better. Along the way, I explain the terminology used by traditional phone companies. By chapter's end, you'll know about all the charges billed by traditional phone companies — as well as which of those charges your conversion to VoIP can eliminate or drastically reduce.

Accessing the Network

All phone costs start with leasing some sort of access line or set of access lines from the *local exchange carrier* (affectionately called *LEC,* which rhymes with "heck"). For the line itself, you pay a monthly access fee that varies depending on the type of line you lease. For most consumers, the line is a POTS line that permits them to place and receive telephone calls on the PSTN. Local line access costs the typical residential customer an average of about $25 per month, not counting recurring per-minute usage charges, toll charges, regulated fees, and taxes.

For businesses, regular POTS line access costs two to five times what a residential customer pays. If you run a small business, you might lease a group of lines and accept, as consumers meekly do, the telephone numbers assigned and provided by the LEC. A larger company with an in-house telephone system might lease different types of access lines. Some may be POTS lines that support two-way access, permitting callers to place and receive calls. Other types of access lines may be used depending on the size and type of company. (I'll cover the other types of access lines beginning in Chapter 4.)

You may hear the term *two-way* used to refer to POTS lines that can be used to both make and receive calls from the PSTN.

Never one to make things simple, your LEC has a different monthly access cost for each type of line. In addition, if your company has its own telephone system, you pay a one-time fee to buy a bulk list of usable telephone numbers that you assign to your employees. As employees come and go, your company can reassign those telephone numbers accordingly.

Some companies lease higher bandwidth access lines, which are much more expensive. These lines combine bandwidth and provide what are known as *POTS line equivalencies*. In this way, companies can reduce the total number of physical POTS lines needed and therefore reduce their monthly line-access costs. But to be able to do this, the companies must have their own in-house telephone system, which introduces another cost to the mix. (Telephone systems are introduced in Chapter 2.)

After you establish access, you can make and receive calls to and from the PSTN. That's when recurring usage charges kick in. (Just when you thought you were safe!) Usage charges for consumers are based on two factors:

- Total length of the call in minutes
- Calling service category

Timing the call to calculate your usage charges is pretty simple: Multiply the total time in minutes by the rate per minute. Although this is considered simple math, few people time their calls. This is especially the case when the calls originate at work because we all think "it is a business call and the company pays for it."

The other factor, the calling service category, presents a more complex challenge because few people know or care to know the differences among calling service categories. There are five such categories, four of which relate directly to your toll usage charges each month. A description of each is coming up next.

Service Categories Cost You Big Time

If you've ever tried to read your monthly phone bill, you know that the system of charges for traditional phone services is virtually incomprehensible to the average person. One of the big benefits of VoIP is that it makes the POTS-PSTN model, together with its complicated billing structure and weird terminology, just go away. Traditional carrier services, unlike VoIP telephony, are heavily regulated. That's why you get a phone bill with more small print than the phone book, filled with monthly line-access charges, per-minute usage charges, service charges, taxes, and special fees all applied to the number and type of individual lines your company or family uses.

To add to the confusion, under the PSTN model used in the United States, recurring service charges are tiered into service categories. The service categories were developed over the years by the telecommunications industry along with the Federal Communications Commission (FCC) and the various states' governing authorities.

After your phone company charges you line-access costs and for any call features you add to each line, they bill you for per-minute usage charges based on your service category. Each phone line you use may be billed for any of the following five service-charge categories:

- Local
- Intralata
- Intrastate
- Interstate
- International

Figure 3-1 provides a bull's eye diagram of the five service-charge categories, organized by the degree of regulation. At present, international service is the most highly regulated category. Interstate is the second highest in terms of regulation. Intralata and intrastate come in third in terms of degree of regulation. Local service continues to be the least regulated.

But higher regulation and longer distances don't necessarily mean higher cost anymore. Interstate is more regulated than intrastate and intralata, but it is much cheaper on a per-minute rate basis. For example, a corporate customer calling from Pittsburgh to Los Angeles (longer distance) might pay $.02 per minute. A call from Pittsburgh to Philadelphia (shorter distance), on the other hand, could cost anywhere from $.06 to $.62 per minute.

In the POTS-PSTN way of doing telephony, more regulation translates into more add-on service charges per line. Under VoIP, you can eliminate all regulated fees and charges because VoIP is totally nonregulated.

It is virtually impossible for VoIP to eliminate all charges for phone service. For instance, if you are a consumer, I recommend at least one POTS line in the home for 911 service and other local calls. Local ordinances require businesses to have at least one POTS line for fire control and 911. Your primary goal is to reduce or eliminate the recurring monthly costs with the other four service-charge categories, and your secondary goal is to reduce your local service costs. With VoIP, you can accomplish these goals by making most or all toll-related calls on-net. But until the rest of the world converts to VoIP, you still need some connectivity to the local calling area using POTS.

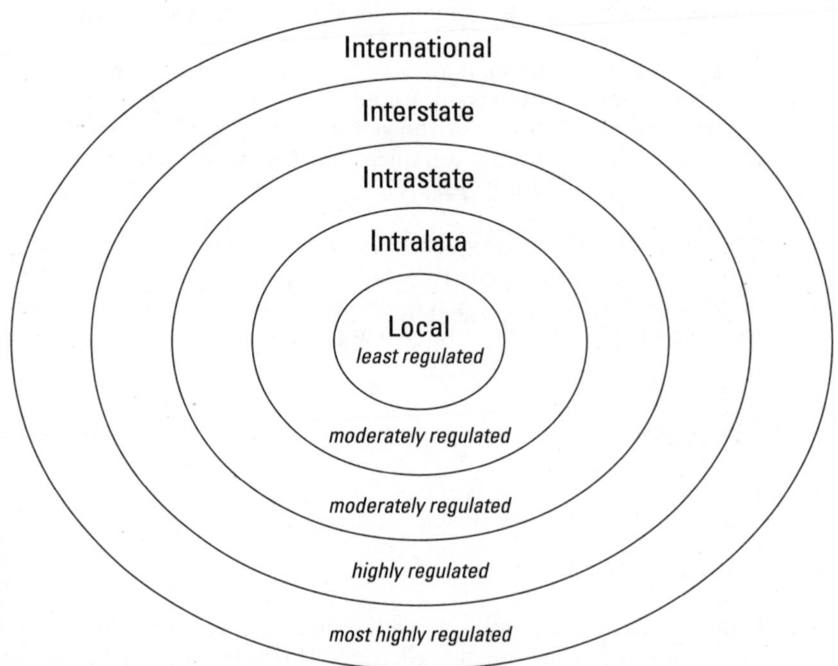

Figure 3-1: A bulls-eye view of regulated charges.

PSTN service categories a.k.a.

The carrier services industry established the five regulated service category names. Of the five, only *local* and *international* have stuck without any a.k.a.'s. The other category names are confusing and unclear to the average person, so other terms have emerged:

- **Intralata:** local toll, intralata toll, regional toll; local long distance, regional calling
- **Intrastate:** in-state, interlata, state toll, in-state calling, long distance (usually incorrect, depending on the context, but frequently used)
- **Interstate:** long distance, LD, state-to-state, toll-calling services, calling across state lines

Paying the local piper

So you went out and got yourself a local-access line for your home or a slew of access lines for your business. Just how much do you pay for the local calls you place on those lines? Figuring out those costs is a little complicated.

The *local* rate category refers to the immediate geographical area, usually no more than a one- to two-mile radius from the telephone from which you're placing a call. Your LEC may try to make the local calling service appear to be free, but the fees are built into the monthly line charges.

Then there's the tricky bit about what exactly constitutes a local call. The LEC's customer-service people can readily specify what your local calling area includes, if you ask. But they are not required to make you understand. You have to ask them to put it in writing or refer you to existing documentation that defines the local calling area for your area. Also, if you live in a larger metropolitan area (such as New York, Pittsburgh, or Dallas), the directories provided by your LEC have maps that indicate what areas and prefixes make up the local calling area.

Why do you need to know your local calling area, anyway? Because not knowing can cost you big time. Today, if you're a residential customer, most LECs provide free local calling service. The LEC calls this *unlimited local calling.* People tend to interpret this as unlimited free calling within their local area. But you want to know a secret? LECs earn the greatest revenue on local calls that terminate in the local toll, regional, or in-state toll calling areas.

Most callers do not know the difference between *local unlimited calling* areas and *local toll calling* areas. The term *local toll* is ambiguous (and should be outlawed). As a result, customers often think they're making free, local calls when in fact they're paying recurring minute usage charges for what amounts to intralata or intrastate calls. Because these "hidden" charges don't turn up until their next monthly bill, they may not realize this is happening until it is too late. Because the LEC is making more money from this lack of awareness, don't hold your breath waiting for them to clarify it for you.

For example, from my Pittsburgh downtown residence, I can call anywhere in the immediate downtown area for free, and each call in this local area can have an unlimited amount of minute usage with no extra cost to me. But whenever I make a call to my doctor's office in Murrysville, a Pittsburgh suburb 19 miles away, I am charged $.12 per minute. Hence the puzzling term *local long distance.*

Do not be misled by all this unlimited, free, local call mumbo-jumbo. If your local service is free and unlimited, it is because the LEC has every hope that you will make lots of calls just outside the local area. Then they can hit you with intralata toll charges and laugh about it all the way to the bank.

If you're a business, the problem is even more acute and costly. Businesses must pay recurring charges for calls in the local calling area. If you work in a business with multiple lines, every PSTN outbound call placed on your business telephone is charged a variable per-minute rate based on the destination of the call. In the Pittsburgh region, for example, businesses often pay $.05 per minute for calls in the local calling area. If you add the total cost of each POTS line to the total minute charges for all the local calls made on each POTS line in the company within a given month, you can begin to see exactly how much you're dishing out for what you thought was a free local area call.

Even if your company uses higher bandwidth transport lines (covered beginning in Chapter 4) with an in-house telephone system, some LECs consider local calls from this type of line to be running on POTS-equivalent lines. With that little semantic sleight of hand, they still charge you the local recurring usage charge on a per-line basis.

Going the distance with intralata rates

Intralata refers to calls that terminate outside the local calling area but within your local access and transport area (LATA). Unfortunately, most people don't know the boundaries of either their local calling area or their LATA. If they did know, they could manage their intralata calls and charges much better.

It used to be that if you leased an access line from the LEC, they automatically became your toll-call carrier. Today, customers can choose what carrier they want for calls made outside the local calling area. If you don't select an intralata carrier when you begin the lease of your local access line, though, you automatically inherit the LEC as your intralata carrier.

Say you're in Pittsburgh and are calling your son, who is at school at Penn State University's local satellite campus eleven miles away. Such a call crosses local calling areas within the Pittsburgh LATA, making it a call in the intralata service category.

First the call goes into your carrier's nearest switching facility. Based on the area code and prefix of the number you dialed, the call gets routed over the PSTN to the carrier's destination switching facility. From there, the call is sent over the PSTN to the facility and switch that terminates your son's line. Then the call goes to your son's line and the actual telephone attached to his line in his dorm room. The telephone rings, and when your son answers the phone, a session is established for the call's duration. As a consumer, you pay a hefty per-minute charge for this type of intralata call.

Intralata calling for consumers is expensive in comparison to local calling rates. Also, consumers have little choice in intralata calling plans. It's always a per-minute rate that can change as the distance between endpoints involved in the call increases or the availability of carrier facilities increases outside the local calling area. If you or your company plan to stay on POTS-PSTN telephony, ask your carrier to give you (in writing) the boundaries for each of the five regulated service categories.

For example, the call to my son is a higher rate than the call to my doctor in a suburb of Pittsburgh, even though this suburb is farther away. Why? The call to my son must travel over rural areas where the carrier has less up-to-date facilities and must use more expensive, slower, alternate routes.

But both calls would usually be less than a call to Philadelphia, which is still in the state of Pennsylvania but farther away from my local calling area than either my doctor or my son's campus near Pittsburgh. The cost of a POTS call usually increases for the consumer based on the distance of the call — outside the LATA. For a business using POTS, it may increase, be a flat rate, or be free, depending on what type of service level agreement exists between the business and the carrier.

Intralata calls are carried the same way for consumers and businesses, but businesses can negotiate intralata rates with their regional (that is, intralata) toll carrier. (Consumers can merely select a carrier and live with the per-minute rate they are assigned.) The carrier can offer businesses a bundled deal based on the anticipated yearly volume of minutes and the amount of minutes the company is willing to commit to.

Carriers can offer business customers flat nonrecurring rates. For example, they can offer a flat rate per call with no recurring minute charges. In this way, the carrier is selling intrastate carrier services like the LEC sells local carrier services. Sometimes you can fashion a long-term deal based on the monthly volume of call minutes. Because carriers are eager to get longer-term contracts, they make those deals look very attractive. But don't be misled: Rates and call volumes change over time. If you're locked into a long-term deal, it may not look like such a deal after several rate hikes.

If your company fails to reach the projected volume it commits to in a monthly, yearly, or even longer-term deal, you'll usually incur penalties. Make sure you check the fine print on any agreement.

How can VoIP help you with intralata charges? It varies depending on whether you are a consumer or a business. If you are a consumer, you would enter into an agreement with a VoIP carrier. You would pay a flat charge per month to call anywhere outside your local calling area.

If you are a company with multiple locations in and around one or more LATAs and your locations are connected with a VoIP network, all the calling that goes on between these various locations is free of recurring carrier service charges. When one location needs to call someone off-net at a distant location, the call can be carried on-net as far as possible before it goes off-net.

Intrastate service rates

The next service category is *intrastate,* which involves carrier services for calls outside the LATA but inside the boundaries of the state where your local access line is installed.

As with intralata, if you tell your LEC nothing about which intrastate carrier you want to use when you begin the lease of your local access line, you automatically inherit the LEC as your intrastate carrier. Intrastate services are basically the same as intralata services except they cover a much larger geographic area. Intrastate is sometimes called interlata because several LATAs are situated in any given state.

If you're a consumer, you're probably paying an intrastate per-minute rate for all your in-state calls with destinations outside your specific LATA. If you're a business, your carrier services company probably set up some kind of plan based on a flat rate with a certain minute-volume or number-of-calls commitment level.

Carriers that own their telecommunications network infrastructure are better equipped to offer you bargains on in-state calling if you sign a long-term deal with them for both local and intrastate carrier services. Carriers that lease lines from larger carriers and then resell carrier services to you have less flexibility.

Many carriers offer a flat rate for in-state calling. In Pennsylvania, for example, some offer flat rates with no recurring charges. But if you're not using VoIP, read the terms of the deal carefully. Does it start out as a recurring per-minute charge and then metamorphize into a flat rate only after you've used (and paid for) a certain number of minutes?

Planned confusion

Intralata and intrastate are among the most misunderstood rate categories. It's no surprise that LECs want you to stay confused. Companies with multiple locations across different LATAs are the hardest hit by charges in these two categories of carrier services. I've worked with some clients who rationalize these charges as normal business operating expenses that they can write off. That's true, but you're still throwing money out the window. Newer technologies such as VoIP can make mincemeat out of intralata and intrastate recurring charges. If you're not ready to use VoIP yet, at least be careful when it's time to sign on with a new intralata or intrastate toll carrier.

So how can VoIP help? VoIP has no recurring carrier service charge for calls to locations around the state that are on your company's computer network, such as your branch office up north or your factory down south. Also, if you have locations around the state on a VoIP network, you can place a VoIP call through the location nearest to your calling destination. Your call would then usually become a local call in that location's calling area.

Interstate carrier service

Like Dante's circles of hell, phone carrier services just keep spreading out. The next circle of charges is interstate. *Interstate* includes calls to a destination outside the local calling area's state but still inside the United States. Interstate is sometimes referred to as calling across state lines or state-to-state calling.

More often than not, interstate calls involve more than one carrier. For this reason, it is difficult for businesses (or consumers) to get any special deals based on their usage. If your carrier doesn't have facilities covering all these types of calls from the point of origin to the calling destination, they lease services from other carriers. These costs are passed on to the carrier's customers (that's you).

How can VoIP help you with interstate charges? For consumers it is basically the same process as with intralata or intrastate costs. You pay a flat charge per month to a VoIP carrier. These carriers have different levels of service, but most permit you to make unlimited calls anywhere in the country for no additional cost.

If you're a company that has a VoIP network spreading across multiple states, all the calling that goes on between the various locations is free of recurring interstate carrier service charges and regulatory fees.

Slamming

One time, my son Gabe, then age seven, answered the phone. Someone asked, "Is your mother home?" Gabe said "yes," and the phone went dead. We didn't know what to make of it. However, on our next month's telephone bill, we saw that our interstate toll carrier service had been changed.

This happened in the era of slamming. Between 1996 and 2000, unscrupulous toll service carrier salespeople would call consumers and trick them into saying the word "yes." After getting the "yes," these salespeople would hang up and cut an order to have that consumer's toll services changed over to their company's carrier services. When the LEC would question the toll carrier's service change request for that customer, the toll carrier would state that the customer said "yes."

Now, carriers are no longer permitted to get away with such unethical practices, and you can safely change over to any long-distance carrier you prefer and not worry about others trying to switch you against your will. But turnabout is fair play, and these same companies that slammed you at your dinner hour are now facing a tough future. Over the next few years, VoIP will probably eliminate the need for these carriers or at least cause them to switch over to a more VoIP-oriented line of service.

International carrier service

Last but not least is the service category known as international. *International* service originates where you are and terminates in another country. This is the carrier service category that is ripest for elimination by VoIP networks because international carrier service is the most expensive per minute of the five regulated categories. Among corporations that do lots of international calling, it's no surprise that a movement is building to deploy VoIP.

Remember our old friend regulation? Much of the cost of international calls comes in the form of increased regulatory fees. In fact, more recurring regulatory costs are associated with international minutes than with any of the other four categories of regulated service.

Companies can run a VoIP network globally. They can run a VoIP network for domestic calls. Either of these network approaches can also have an attached VPN (virtual private network). The VPN can support telephony calls coming in or going out through the Internet. These network options would eliminate most of the cost of international carrier services calling. Kind of makes you want to open a branch office in Paris or Beijing.

VPNs use the carrier transport capabilities of the global Internet to support mobile and global communications. VPNs started out supporting computer data and e-mail back in the mid-1990s. Since then, VoIP has added telephony to the list of VPN applications. I discuss VPNs in greater detail in Chapter 9.

If your company doesn't make many international calls, VoIP can still save you a barrel of money in all the other domestic-based rate categories.

Summing up carrier services

Several layers of costs exist whenever you make a phone call due to regulations and the categories of carrier service. All five categories combine to form the bulk of the monthly recurring charges for customers (both residential and corporate) under the existing PSTN model of telephony.

Carrier service companies do just about anything to keep their corporate customers using their services. They try to make a strong business case for your company to stick with them. If you're unaware of the different rate categories, your company's total number of access lines, and how each category and line relates to your company's particular telephony needs, it can cost you big bucks each and every month.

Your carrier may try to reduce your per-minute rates across the board. In the case of intrastate calling, they may offer flat rates that are not metered and charged by the minute. However, the carrier wants some kind of commitment from your company in return. The commitment can usually take one of two forms.

The first form is a volume-of-minutes plan. In this form of the plan, your company, including all the locations connected through the carrier's network, agree to use a specific, aggregate number of minutes on their carrier network. In this form, they always specify a term within which the minutes must be used, such as ten million minutes per month or per year. Read the small print regarding what penalties may apply if the company fails to meet the volume-minute quota in a given month or year.

The other form may relate to the total number of calls made irrespective of the total aggregate volume of minutes across your company's enterprise. You can see this form when the company already has a flat-rate charging plan in place. For example, instead of paying metered charges for local, intralata or intrastate calls, a company might have a flat rate such as $.05 or $.06 per call. The recurring minutes or total usage minutes become secondary as a cost factor — with flat rates, the number of instances of placing calls are used to figure your carrier service usage bill.

It's also possible to have a plan that combines both minute-volume and flat-rate plans into a sort of hybrid plan. The larger your company and the more locations in different states, the more you'll need a thorough analysis to achieve the optimal plan if your company's telephony needs are to be satisfied by a POTS-PSTN carrier.

The biggest pitch that carriers use to get your signature on a multiyear service contract is to promise you deep discounts based on your company's overall calling volume in minutes. For example, if your company does several millions of intralata minutes per month across all your company's locations and you're paying an average of $.07 per minute, the carrier's account representative might offer you a new deal that reduces your intralata costs to $.05 or $.06 per minute.

After you're using millions of minutes, a $.01 change in the rate translates into a lot of money. For instance, one million minutes at $.01 per minute equals $10,000 of cost to your company in one month. That kind of savings would sound swell — if you didn't know that under a VoIP network plan you would have little or no charges for intralata carrier services. In a VoIP network, all on-net traffic would cost $0, and any calls that must go into the POTS-PSTN network would be reduced to local calls.

Your carrier account rep won't want to tell you about the penalty if your company fails to meet the volume commitment in any given month. The penalty could be an even higher per-minute rate than you had before the new deal or an increase in the term of your contract by one month for every month that you fail to meet the minimum.

If your company must stay on the POTS-PSTN carrier network, I suggest that you consider evaluating VoIP, if only for a few telephones or one small local area network at one of your site locations. If you already have the LAN running, your cost will be minimal. Let your POTS-PSTN carrier account rep know that that you're looking at VoIP, and see how fast it gets him or her to come around with a new deal that seriously reduces your monthly carrier charges. But don't sign a long-term deal; as soon as you complete your VoIP testing, you'll want to put much of your company's telephony on VoIP.

Saving with VoIP

If you've read the chapter up to this point, you're a much more savvy POTS-PSTN customer. You now know exactly how your carrier makes money at your expense. You also know how the five regulated service categories can combine to increase your monthly and annual telephony costs and therefore reduce your revenue. Something that increases costs and reduces revenue is something you need to control or change. VoIP can help you do exactly that. So, how will you fare under a VoIP system?

Good news for the family budget

Most carriers no longer apply service charges to residential accounts for local calls, beyond the monthly recurring costs of the access line. All other rate categories are billed on a per-minute basis. If you convert to VoIP for your home phone, you typically have no recurring service charges for any of the other services categories for calls outside your local area. This, in itself, is a tremendous savings for anybody who does even a moderate amount of nonlocal calling.

Because you must use broadband to get VoIP in your home, I suggest that you pay the additional fee to keep your POTS phone connected to your broadband service. Use the POTS phone for local calls and 911. When the older PSTN catches up with the newer VoIP technology and can support E911 (enhanced 911) calling, you can drop the POTS connection. (With E911, your contact and address information is transmitted along with your call to the 911 emergency center.)

Taking savings to the office

If you run a business, local-area calling-plan charges average about $.05 per minute. Under a VoIP model, the cost of calls to the local calling area are the only significant recurring usage charges you won't get rid of — at least not until the rest of the world adopts VoIP.

Even local carrier charges can be reduced under a VoIP model, however. You can do this if you leverage volume by total minutes and make a contractual commitment to the local carrier. Tell them you are going VoIP, and see how quickly they will accommodate you. If most of your local calls are to other offices on your company's network, VoIP eliminates any recurring service charges for those calls because this traffic is on-net.

It's in the other service categories — intralata, intrastate, interstate, and international — where your company can save the bulk of the usual monthly service charges by using VoIP. Keep in mind that these monthly charges can be huge. One of my clients had 367 locations across the country and 17 international locations, and a combined computer and telephone network billing of $4.2 million per month. And 75 percent of the billings were telephony carrier services charges. That's about $3.78 million per month for POTS-PSTN telephone services. VoIP would eliminate more than 90 percent of these telephony charges because the company is already paying for its computer network. Under a VoIP telephony model, any company with substantial intralata, intrastate, interstate, or international calling service requirements saves a bundle of cash.

Toll-bypass: Saving with calls at a distance

The same VoIP cost-savings rationale for the international company described in the preceding section applies to intralata, intrastate, and interstate calling carried on-net over a company's VoIP network. Although the costs on a per-minute basis for interstate have come down significantly since the Telecommunications Act of 1996, a company with many locations across many states can accumulate millions of minutes per month just to support the communication that goes on among all its locations.

Did you ever take a plane trip only to find out you paid $500 more for your ticket than the person next to you? Interstate per-minute charges for businesses today make airline ticket pricing policies look downright logical: These charges run the gamut from less than $.01 to $.10 per minute. Interstate charge plans for companies using the PSTN vary so much because the plans depend on the minute-volume commitment and the plan's term. The longer the term a company agrees to, the better the current rate provided by the carrier.

However, under a VoIP approach, your on-net calls have no carrier service charges. Therefore, you have no need for a telephony carrier services price-term plan. Also, on-net calls eliminate the toll charges that come with all the various calls to areas outside the local calling area. Because the calls travel over your VoIP network, you don't use the LEC's facilities. A company's monthly carrier service charges for all on-net interstate calls made on the company's VoIP network total nada, nothing, zip, zero. You get the idea! This benefit has become known as *toll-bypass.*

POTS-PSTN per-minute calling costs remain highest for international calling. These costs can be eliminated or greatly reduced if international calls are carried over your company's private VoIP network. Much of the cost of an international call comes from the huge regulatory fees that pile up as the call moves along from country to country. But a VoIP network eliminates all those fees. In addition, VoIP makes these recurring charges your favorite number and mine: $0.

Add-on recurring costs

Tallying the costs of traditional phone service is like adding up the cost of sending your kid through college. There just doesn't seem to be an end in sight to the charges. As if the access line costs and recurring carrier service charges weren't enough, you must deal with other monthly costs and regulatory fees. These payments, which go to various government entities rather than to your LEC, are based on a percentage of each line's monthly access cost. Examples include the Federal line surcharge and the 911 fee.

It may seem like these monthly charges are nominal but, just like the national debt, they really add up. Just dig out your last phone bill and take a look at the total cost. Depending on the location of your telephone lines (that is, which LATA applies), these regulated fees typically total about 4 to 7 percent of your total monthly access costs. A medium-size company with twenty locations, call centers at each location, and lots of calls across the United States racks up a monthly carrier services bill of approximately $500,000. This company would be looking at add-on monthly recurring costs of approximately $20,000 to $35,000. That annualizes out to $240,000 to $420,000. Final costs depend on the LATAs involved and the specific types of access lines. But these are add-on costs that do not need to be counted in a VoIP network; they largely go away.

If you convert to VoIP, you're still charged regulatory fees for your dedicated network lines, but you already pay these costs to support your computer data network. You do not have to pay them again because VoIP calls are carried on your computer network. There are no additional regulatory costs for running VoIP telephony over your computer network. And with your telephony carrier services needs now being supported by your computer network lines, you can drastically reduce or eliminate the number and types of lines your company needs to support the POTS-PSTN way of doing telephony.

Finally, we come to the calling features, such as voice mail, call waiting, and call forwarding. The LEC charges for features à la carte, and if you've ever ordered à la carte in a restaurant, you know it costs more. This is because you pay for each individual feature (item) separately. For example, call forwarding might be a $5 per month per line charge on top of all the other charges you pay.

Your LEC may be able to bundle features and leverage your company's total monthly usage minutes for all your lines to offer you calling features at a lower cost. Be careful if the carrier asks you to commit to a more lengthy term to achieve cost reductions; in the end, they may cost you more!

Most companies with their own internal telephone system provide their own calling features. With pure POTS and Centrex line models, calling-feature costs can have a big effect on your company's monthly telephone bill. Remember that features are priced based on each line. If your company has hundreds of lines, the overall cost for all features for all lines can be astronomical. For example, adding voice mail ($8), call forwarding ($5), and conference call ($4) features to two hundred lines would cost your company an additional $3400 each month. Wouldn't you rather hire additional employees or install a large-screen TV in the break room with that money? As you may have surmised by now, VoIP comes with all the features of the POTS world, plus many new and exciting ones. (VoIP call features are covered in Chapter 10.)

The bottom line? With VoIP, you can reduce your monthly recurring charges by as much as 95 percent — and that's a lot of money in anyone's book!

VoIP Savings: A Case Study

One of my clients in the Pittsburgh area has eleven locations distributed across several local calling areas within two Pittsburgh LATAs. Five locations are in the city itself. The other six are in the South Hills, with two inside Allegheny County but outside the city, and four located to the south across the county line in Washington County. The client spent enormous amounts of money on phone service because an interoffice call between locations often crossed intralata boundaries.

This company had a patchwork of standalone LANs at each location and a few Internet dialup accounts. Each of their largest two locations had its own phone system, but they defeated part of the potential benefit of those systems by running POTS access lines into them instead of higher bandwidth access lines.

Moreover, the client had many additional access lines that did not terminate at their own telephone system. They leased these lines like a consumer would lease a POTS line, but they were paying business prices and did not connect these lines to their telephone systems. The other nine locations had basic POTS access lines. All told, ninety-one POTS access lines ran across the eleven sites.

In addition, they had two LECs providing their access and eight toll carrier service providers: The client received monthly bills from ten companies! The client's key people were stressed out just from all the bills they were getting. They also couldn't understand from the bills why they had such high charges. Some of the locations were less than ten miles apart but had the highest recurring charges each month.

Analyzing the client's usage

When I came on the scene, I analyzed their monthly billings for the past three months. I found that their total average monthly billings for intralata services came to just under a whopping $11,000 per month, or a projected annualized billing of $130,000. Intralata recurring charges were about 63 percent of their total monthly telecommunications bill.

At our first meeting, I passed out a spreadsheet which detailed their current costs and compared them with the probable costs associated with my recommended solution. See Figure 3-2.

Before we began to review the spreadsheet, I asked, "What do you see as your number one challenge?" Their answer? Customers complained that they got busy signals every time they called. Their second most important concern was that only one of their eleven sites had high-speed Internet access.

Five sites used dialup Internet access, but this service did not come close to what their customers and suppliers thought they should have. Also, when the sites that did have dialup were online, no one could call in through the POTS lines. You would think that for what these folks were paying they would have had the best telephony service and Internet access money could buy. Instead, they were spinning out of control, and they were ready to listen.

Figure 3-2: Monthly recurring and access charges before VoIP.

POTS access lines (72 @ $65)	$4,680
PSTN usage (not including Intralata)	$690
PSTN intralata usage	$10,800
T1 Internet (line and access)	$995
TOTAL	$17,165

The VoIP solution

I designed a VoIP network that provided a dedicated access line between the two main locations. These sites were seven miles apart but in different counties and therefore different LATAs. We put in digital subscriber line (DSL) access at the other nine locations.

DSL was a no-brainer because these sites already had at least two access lines. Because local ordinances require that all businesses have at least one POTS line for emergencies and fire control, we used one of the existing POTS lines for that requirement with no additional cost. Also, because DSL requires that you to have an existing POTS line in operation, it was simply a matter of having the LEC upgrade one POTS line at each location to include broadband DSL service. Figure 3-3 shows their VoIP network.

We selected one carrier for all local and toll-related carrier services. (You could almost hear the sigh of gratitude just for eliminating all the different monthly bills.) Before VoIP, their combined monthly recurring access and usage charges averaged more than $17,000. In the first month of operation under VoIP, these charges dropped to just over $2100. Most of this savings resulted from reduced intralata costs and the elimination of access lines.

The dedicated T1 line enabled us to provide a high-bandwidth private-access link that bridged the intralata boundary between their two main sites. The private line made the two intralata areas one.

Calls originating from any of the sites in the city destined for any of the sites outside the city were routed to the main site within the city, transported over the private VoIP line to the out-of-city main site, and forwarded to the destination telephone. As a result, all on-net calls were treated like local calls for billing and bypassed the regulated intralata charges.

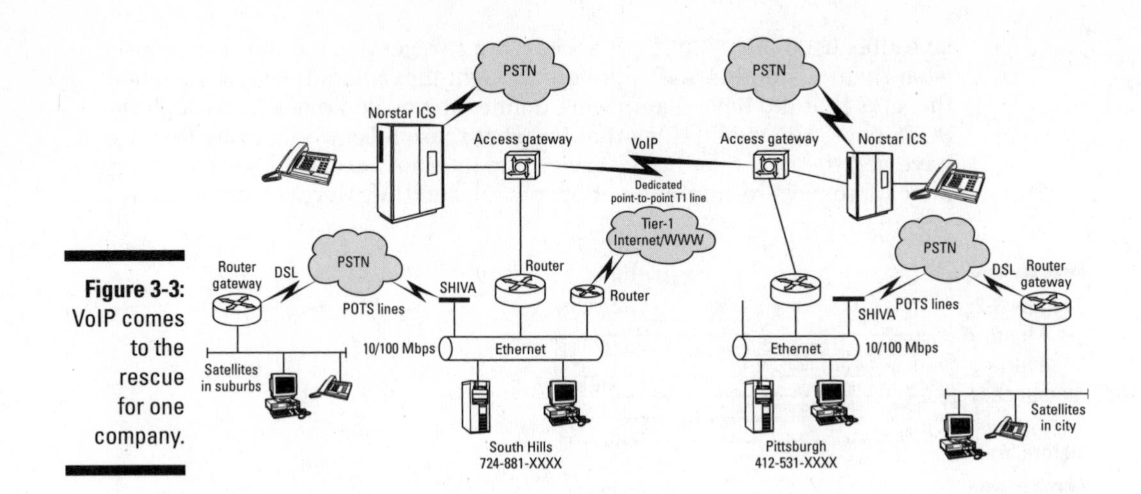

Figure 3-3: VoIP comes to the rescue for one company.

As mentioned, the main site in the South Hills had Internet service, but none of the other ten sites could access it. In the second month, we used some of the money they were saving to put in an Internet gateway in the South Hills. Now every location could access the Internet through the company's VoIP network.

By the second month of implementing VoIP, their intralata toll service charges had dropped to $184 because all interoffice voice traffic was now carried on their private VoIP network instead of the regulated PSTN. Moreover, they could now understand the single bill that covered their entire VoIP network services; this alone brought sanity to their operation. In addition, all eleven locations could now access the Internet. They were already beginning to plan their Web site design and working out processes with their suppliers to use several e-commerce applications that were not possible before VoIP.

Some startup costs were not exactly inexpensive. However, their savings from VoIP more than covered these costs with money left over. Figure 3-4 is a breakdown of what it cost to bring VoIP to their enterprise.

Applying VoIP to your situation

The moral of this story? If your company has facilities distributed across more than one local calling area within the same LATA, look at your intralata toll charges. If all your sites switched to VoIP, you'd eliminate most, if not all, intralata charges you currently incur.

One-time startup charges	
2 IP access gateways	$6,000
2 DS1 (T1) interface cards	$5,000
2 routers w/gateway and Ethernet interface	$6,000
TOTAL	$17,000
Monthly recurring charges	
POTS access lines (22 @ $65)	$1,430
T1 line	$481
DSL service (9 @ $79)	$711
Internet access, main site	$995
PSTN service (not including intralata)	$2,646
PSTN intralata service	$184
TOTAL	$6,447

Figure 3-4: Monthly startup and recurring charges after VoIP.

In addition, you may be able to add services you never thought possible and pay for the services through your savings. In the case of my client, the VoIP network saved them more than $9000 per month even after all startup costs were paid. In addition, they expanded dedicated Internet access to all their locations, upgraded their telephone systems to support VoIP, and acquired needed equipment to make the changes. At the end of the first year under VoIP, their annualized savings were just under $112,000. In short, they got greater service and spent a lot less money. Figure 3-5 summarizes the costs and savings over the first year for this VoIP conversion project.

Recurring costs pre-VoIP	
Monthly access and usage	$17,165
Annualized total cost	$205,980
VoIP Start-up Costs (1-time)	$17,000
Recurring costs post-VoIP	
Monthly access and usage	$6,447
Annualized total cost	$77,364
Startup costs recovery period	2 months
First year savings after costs	
Monthly	$9,301
Annualized	$111,616

Figure 3-5: Summary of annual costs and savings after VoIP.

Do you already have a computer network connecting all your sites, or can you readily upgrade to connect all sites? If so, the cost of going to VoIP is greatly reduced because VoIP operates on the same network as your computer data.

Part II
Taking VoIP to Your Network

The 5th Wave
By Rich Tennant
GAREN HELPS DEVELOP THE FIRST VOICE OVER TURNTABLE PROTOCOL
© RICHTENNANT

"Hello?! Hello, Phillip?! You're breaking up! Listen, put a penny on the tone arm and turn the speed up to 45 RPM!"

In this part . . .

Get ready for a mind-bending tour of how you can integrate voice with your data network through the magic of VoIP. In this part, you discover an expansive view of networking, and how you can take advantage of all the benefits that VoIP has to offer.

Chapter 4 provides a road map to the different methods through which VoIP can be implemented. The chapters that follow look at virtually every method you can imagine: circuit-switched networks, broadband networks, dedicated networks, wireless networks, and the Internet. Each chapter provides all the details you need to determine which network type is right for you.

Finally, Chapter 10 discloses how to use telephones with VoIP. You find out about the latest VoIP-enabled phones, and also see whether you can use your existing phones with VoIP.

Chapter 4

Road Map to VoIP Transports and Services

In This Chapter

- Recognizing the abstract world of CSIs
- Understanding transports
- Uncovering the five carrier service infrastructures
- Fitting VoIP and the Net in the CSI structure

Let's face it — telecommunications can be daunting to those who have not given much thought as to how their voice gets from their phone's handset to their Aunt Matilda in Dubuque. Be that as it may, the technology between you and Aunt Matilda is simply amazing.

This chapter introduces you to the wonderful world of networks, transports, and transport services. Here you discover what a CSI is (besides a great family of television shows) and why you should even care. Before you are finished with this chapter, you'll have a good grasp of things you didn't even know you needed to grasp. (Spooky, huh?)

This chapter lays the groundwork for Chapters 5 through 8. Here you find the conceptual framework that lets you make sense of different ways of transmitting data, such as broadband, dedicated lines, and cable. You also discover why packet-switched communications methods are much better than traditional circuit-switched methods.

If you don't want to take a peek behind the curtain of telecommunications, you don't really need to read this chapter — and you don't even need to feel guilty about it! That's not an effort to discount the information provided here, but a recognition that the information may not interest everyone. You can safely skip this chapter and come back to it later when you finally develop a healthy curiosity as to why things work the way they do in the telecom world.

CSI: Telephony

All types of networks operate within a much larger structure known in the telecommunications industry as a carrier services infrastructure (CSI). The carrier services infrastructure is an abstract concept for most people. As you begin to discover the various VoIP network types (such as DSL or a T1 line), it's essential to know about each network type's underlying CSI. Whatever network type you may choose to use for your VoIP, it is always a subnetwork of a larger CSI.

Figure 4-1 identifies the CSIs. We get into the details of each CSI later. For now, just focus on the fact that there are five CSIs through which all public and private communications travel.

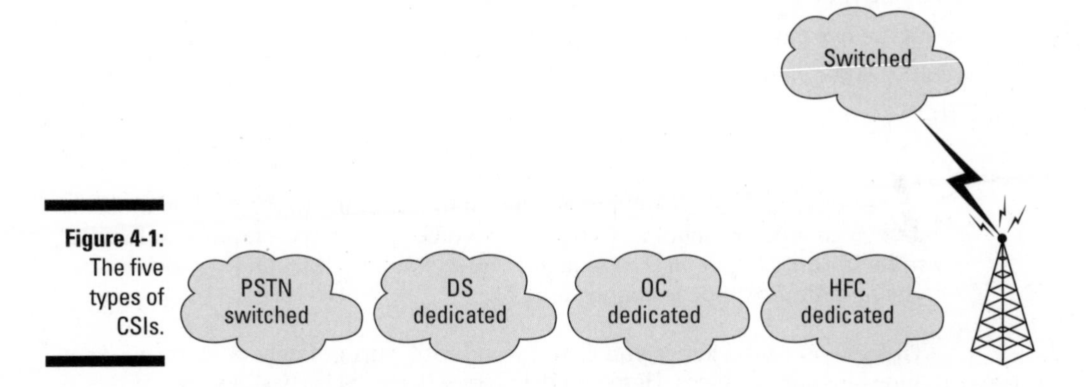

Figure 4-1: The five types of CSIs.

The different CSIs are theoretically owned by the carrier companies that lease the various network transports and services. But in reality you can't own something that exists largely in abstract terms. A good analogy is the National Football League. Is it owned by the NFL? We could say that it is. But to be more correct, we might say that it's owned by all the teams that make up the NFL. And an individual NFL team exists by virtue of the NFL granting (selling and approving) a team franchise.

Truth be told, the NFL doesn't own much of anything in the physical sense of the word. Football fields are owned by their respective teams. NFL players are said to be owned or at least under contract by their specific team.

At the same time, the NFL is not totally out of the picture. We hear lots about the NFL regarding regulatory measures and enforcement actions they may take against teams and players in the NFL. The NFL sets the rules for all teams to operate collectively. All the football stadiums together, with the players and the games they play, form the NFL.

Like the NFL, CSIs are not owned by any specific carrier company. Each carrier has a certain amount of physical network transports (lines) within one or more of the CSIs. Today, more than a thousand different carriers operate in the domestic United States. Many are local and regional. Others are national and even international. Like the various NFL teams, the carriers compete with other carrier companies for business from the corporate marketplace as well as the consumer marketplace. They do this by leasing transports and network services from a given CSI to their customers. Everyone needs carrier services and most companies have a diverse set of telecommunications needs.

All five of the CSIs relate to the telecommunications industry, but each CSI contains different types of network lines and services. In the case of the wireless CSI, which ultimately uses lines at its core, there are no lines for the customer in the physical sense of the word. But there is a frequency spectrum and frequency channels, just like the various station channels that operate on radio.

It pays for you to know your VoIP network options across all five CSIs, but don't expect your carrier to keep you up-to-date. If you're near your contract renewal anniversary, you most likely will hear from several carriers about the latest and greatest services now available. If you're currently using a nonVoIP carrier company, your carrier won't have much of an incentive to talk to you about VoIP because it significantly reduces your dependence on conventional telephony networks.

Choosing a Transport

CSIs are made up of network transports (lines or channels) and the carrier's equipment used to terminate these lines. Each CSI has multiple carrier transports to offer any customer.

A transport is a physical or wireless channel (or aggregate of contiguous channels) that supports the transmission of electrical, optical data, telemetric data, voice, or video signals.

Wow! That's a mouthful. It really is easier to just think of a transport as a physical line. In the case of wireless communications, the transport is a channel that corresponds to a unique frequency assigned to your cell phone or PDA. Currently, more than fifteen transport types are available.

With so many transports to choose from, how is anyone supposed to decide which to use? Cost and bandwidth become large factors in any CSI or transport-related decision. This is true whether you're a large company with offices all across the country, a single location with a large campus-like setting of many buildings, a single location with one or more floors, or just a residential customer looking to reduce your monthly telephone bill and improve your Internet access. Ultimately, deciding what transport to use becomes the key decision point for any network, including VoIP.

To better understand transports, consider the T1 transport, easily the most popular dedicated transport in the corporate sector. It's not unusual for companies to have many T1 lines throughout their private network. One reason that T1 lines are popular is because the bandwidth they provide can be optimized using dynamic channel allocation. This means a T1 line can be subdivided into twenty-four smaller bandwidth channels. These channels may be assigned to a bandwidth pool that supports computer data, VoIP, and video traffic. The terminating equipment for the T1 line can assign, at the moment needed, a specific channel or group of channels to bring up a VoIP telephone call or a videoconference call. When the need is over, the channels are returned to the bandwidth pool.

Dynamic bandwidth allocation is not available on all network transports. This is one of the reasons why it is important to know what type of transport or transports you're planning to use. Your carrier may or may not make these points of distinction in a transport leasing deal. Dedicated (private) transports are not owned by the customer; they're leased and there is a monthly access charge based on the distance from point A to point B. But the higher cost of dedicated transports is well worth it.

An important rule to follow when selecting network transports is to start with the question "What will we be using the network for?" This leads to a discussion of uses for network transports. These applications (uses) are called transport services.

Any transport can be used for one or more transport services. A POTS line, for example, is used for telephone services. Many customers use a POTS line at home for computer modem dialup services or to support broadband DSL services. As a result, a POTS line running DSL can support VoIP telephony in the home.

A T1 transport is often called private line service because it's dedicated to the customer's use. None of the carrier's other customers or customers of other carriers can share the line. The customer may run all sorts of applications (transport services) on the T1 line. These applications may include computer data, VoIP telephony, and videoconferencing.

VoIP is considered first and foremost a transport service. VoIP runs on many of the same transports that support computer data and video. It uses packet-switching protocols to support telephony services, so it has the potential to eliminate or replace older transports and services that can't integrate data, voice, and video on the same transport.

The carriers of the future need to incorporate VoIP and related products and services into their business offerings. The most successful carriers are those that not only include VoIP but can lease VoIP services out of two or more of the five CSIs.

The Five Golden Rings of CSI

It's useful to understand from which CSI a network transport service comes. The CSI to which a transport service belongs affects the way in which that transport service is implemented. Figure 4-2 shows the five types of CSI, along with popular transports and VoIP transport services.

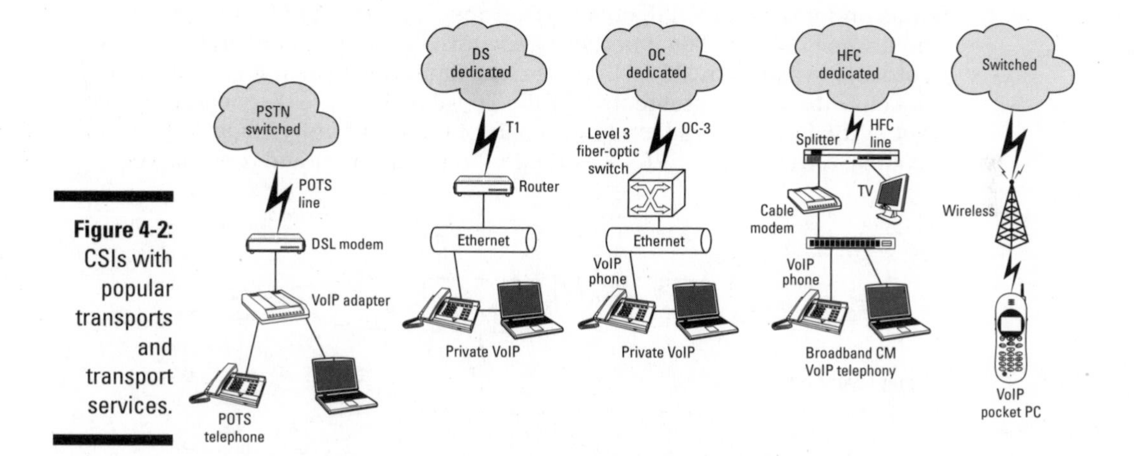

Figure 4-2: CSIs with popular transports and transport services.

As an example of how a CSI and transport can affect a transport service, consider the case of VoIP. An integral part of VoIP is the concept of transmitting voice packets over the transport. But each CSI, each transport, and the VoIP transport service provided therein support voice packets differently. This is because each CSI operates with different underlying protocols.

A transport line on the PSTN CSI, for example, is a limited bandwidth, circuit-switched medium that can support telephone services. A T1 transport is a much higher bandwidth line out of the DS CSI that can support many transport service applications that use packet-switching protocols. Although VoIP works under both the PSTN and DS CSIs, the faster T1 transport on the DS CSI affects the efficiency with which VoIP operates. If you're thinking about using VoIP or have concerns about the operation of your existing VoIP, evaluate your network transports and transport services, paying particular attention to which CSIs are involved.

The following sections examine, in some detail, the various CSIs available, along with how VoIP works in relation to those CSIs.

The PSTN CSI

The public switched telephone network (PSTN) is the oldest CSI, actually beginning with the work of Alexander Graham Bell in the late 1800s. I won't go into too much detail about this CSI's history because I cover it in Chapter 2.

Over the years, carriers have been installing, expanding, improving, and interconnecting various parts of the PSTN. Compared to all other carrier service infrastructures, the PSTN is the largest.

It was under the PSTN CSI, on a POTS transport, that VoIP first started — and it wasn't pretty. The inherent bandwidth limitations and circuit-switched protocols required by POTS impose clear limitations when it comes to the packet-switched requirements of VoIP. These limits were discovered by accident in the first VoIP call made by a pair of Internet hobbyists in 1995. Figure 4-3 illustrates how the PSTN was related to the Internet and the first VoIP call.

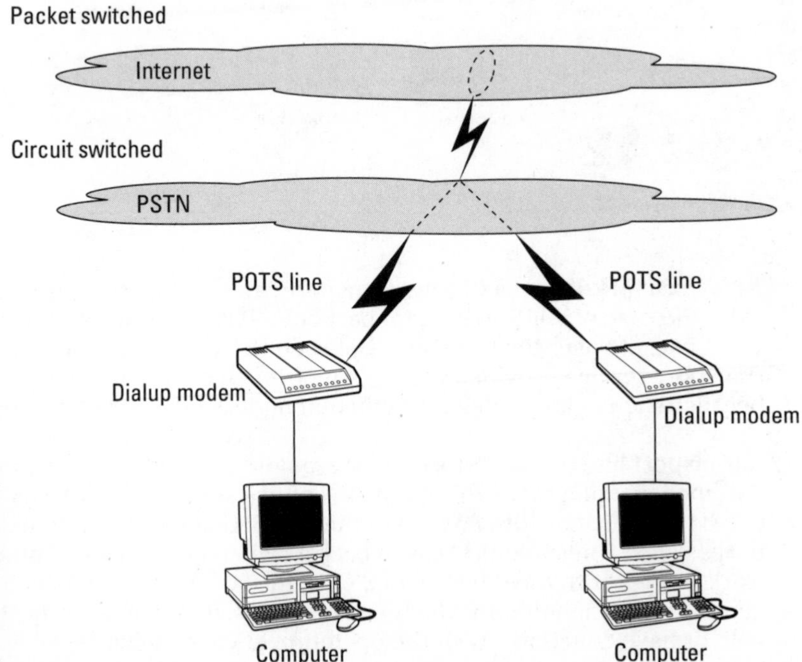

Figure 4-3: The first VoIP call utilized the Internet through the PSTN.

The earliest VoIP experiments were not pretty by today's standards. But by the late 1990s, VoIP was being viewed by dot com startups, the carriers, and even several telecommunications equipment manufacturers as having great potential for the future of packetized telephony.

Even though VoIP can work over the PSTN for a single call, it's not a viable solution for large companies that need to make multiple calls at the same time. Quality of service (QoS) quickly comes into play, and dedicated lines start becoming the minimum level at which sufficient QoS can be achieved. The PSTN CSI doesn't provide dedicated lines, so it doesn't provide a suitable solution for robust VoIP.

Poor QoS in the PSTN CSI is caused by the inherent bandwidth limits of POTS, the circuit-switched protocols used on the PSTN, and the fact that the number of switching hops in a POTS call add too much overhead into each and every packet in the transmission. This overhead, more than anything else, is the major culprit.

That said, if you are an individual (not a company), and you want to run only a single line over VoIP, you can get satisfactory QoS using a regular broadband connection such as DSL. Better still, this makes VoIP quite affordable because the cost of DSL is relatively low these days. However, using such a transport for VoIP is not reasonable for larger businesses that need better QoS for larger numbers of calls.

The DS CSI

In 1964, the carriers began channelizing and aggregating analog inbound telephone calls onto digital, high-bandwidth transports. The digital service (DS) carrier service infrastructure was born. The type of wiring used for DS transports was copper, like the PSTN transport lines in the PSTN. The DS transport lines, however, were of a thicker gauge and were capable of sustaining higher bandwidth capacities. The carrier often referred to DS type transport lines as "high-cap" T1 lines to distinguish them from other types of copper transport lines in the PSTN. Today most T1 transport lines are provided using fiber-optic lines.

Because of higher bandwidth capacities, transports under the DS CSI are terminated differently than lines in the PSTN CSI. This is in part what led to the designation of DS lines as being dedicated. Unlike PSTN transport lines, which potentially could be terminated and switched out all over the CSI prior to connecting to their destination point, DS transport lines were installed to provide a direct connection between source and target destination points.

DS transports do not share any switching points with other customers in the carrier service infrastructure. For this reason, after a DS transport line is installed, it is said to be "nailed up" for that customer's use only. This direct, nailed-up connection enables not only higher bandwidth, but also much greater throughput and no contention from other parts of the CSI.

Contention occurs when a number of users try to use a limited number of resources at the same time. In a public network, all users vie for a limited amount of bandwidth. That's why you get the "We're sorry, but all circuits are busy now" message periodically. A dedicated DS line has no contention because it has no other users — it is your line and your line only.

VoIP transports can be dedicated

When Ethernet LANs were standardized in the late 1980s, a huge demand emerged from big multilocation companies wanting to connect their LANs in a wide area network (WAN). Many had hundreds or thousands of locations, each of which was running its own Ethernet LAN. Back then, networks had strictly a computer data context (VoIP had not yet been discovered).

In response to customer demand, the telecommunications industry provided frame-relay transport services, one of the most important transport services to come out of the DS CSI. Frame relay takes the frames on the LAN side destined for another location on the WAN and packetizes them for transport over the WAN. Today, 86 percent of corporate America continues to use frame-relay for computer data service.

A frame is just another word for a data packet. Technically, a frame is a data packet on a local network. Only when the frame is encapsulated for transfer over nonlocal networks (see Chapter 1) does a frame become correctly referred to as a packet.

Frame relay is losing ground to DSL because DSL is now available at commercial bandwidth levels. But frame relay still appears to be the transport service of choice when it comes to interconnecting large, multilocation, data-only LANs. Usually all sites are connected with a T1 or T3 transport line, but under frame relay they do not always operate with the line's full capacity. Thus, their bandwidth is purposefully throttled by the carrier to provide frame-relay service through what would otherwise be a very large "pipe."

The carrier charges a monthly access fee for the transport line itself. In addition, charges are paid monthly for the transport usage in a frame-relay network. This charge is for port speed, which is based on the number of channels (versus the entire transport line's capacity) that the customer uses. Therefore, it is not uncommon to see a lot of fractional T1 (a T1 line that uses only a fraction of the total twenty-four channels) services in a frame-relay network. The good news is that any frame-relay network can be updated cost-effectively to support a dedicated VoIP network because the T1 or T3 transport lines are already in place.

The DS CSI's two most popular transports are the T1 line, which has 24 DS0 channels, and the T3 line, which has an aggregate capacity of 672 DS0 channels. (DS0 channels are 64 Kbps channels, as described in Chapter 7.) Because DS transports are dedicated and channelizable, the T1 and T3 transports work well with VoIP. On a dedicated transport, specific channels on the DS line can be allocated to VoIP calls when needed and returned to the DS transport's channel pool when not needed. As a result, DS transports can be used not only for VoIP but also for integrated computer data and videoconferencing.

Other VoIP transports

Many companies are finding the T1 line an effective transport for supporting VoIP. The cost of a T1 line has dropped significantly in the past five years. It is still priced based on total mileage covered, but with the emerging fiber glut, many T1 lines can be leased to companies from the carrier's excess fiber-transport lines. When this occurs, the T1 line is said to be carved (multiplexed) out of the much higher bandwidth fiber-optic transport line. A fiber-optic line has enough capacity for thousands of DS0 channels.

As mentioned, a T1 line provides twenty-four DS0 channels. If a fiber-optic transport line is already in place, it's just a matter of the carrier programming their equipment to deliver the twenty-four channels to the customer. Figure 4-4 illustrates how a building containing multiple businesses typically gets its transport access lines. The LEC delivers a single, huge bandwidth pipe, in this case an OC-3. The OC-3 is then subdivided as needed to provide various types of other bandwidth lines.

The LEC often installs a larger transport line and then throttles back what is delivered through the line because the labor costs are about the same for any dedicated line. The LEC's logic is reasonable: Pull (install) the most effective high-bandwidth transport possible. In this way, they position themselves to support the current and future bandwidth needs of all the companies in the building. The LEC expends labor costs once in return for many future bandwidth requests.

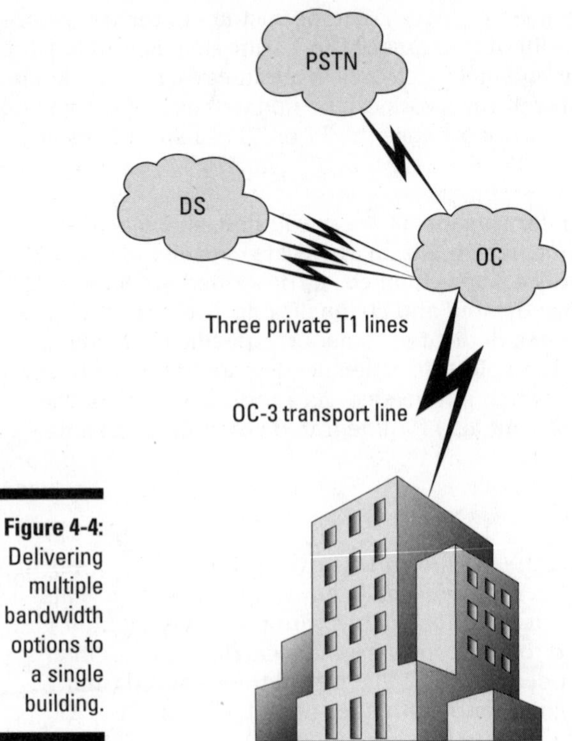

Figure 4-4: Delivering multiple bandwidth options to a single building.

The optical carrier CSI

The DS carrier service infrastructure gave us two important building blocks that were used to further extend the capacity for supporting VoIP networks. First and foremost, the DS network established that analog signals could be regenerated in digital format. Second, the DS network established that digital signals could be aggregated with other digitally regenerated signals in the form of DS0 channels. Thus, the capability to channelize digital bandwidth evolved. Dedicated channels have proven that they can support VoIP with the same if not better quality of service that we have come to expect with POTS over the PSTN.

With the DS series of standards established, a basis existed for specifying how we might further scale and extend bandwidth capacities when the new fiber-optic cabling carrier services infrastructure evolved. Compared to fiber-optic cables, the copper-based wiring of the PSTN and DS CSIs is much more expensive to install and more prone to failure due to electromagnetic interference, weather, and the need to protect the wiring inside expensive conduits.

Fiber-optic cabling uses laser light and is not as vulnerable to these elements. Moreover, fiber-optic cable is more flexible and easier to install. And after the use of fiber-optic cable reached critical mass, it became far less expensive to install compared with nonfiber alternatives.

In 1982, the first fiber-optic cabling systems were commercialized. That same year, MCI became the first telecommunications provider company to choose fiber-optic cable to support its national POTS carrier network. Since the 1980s, an entirely new, totally fiber-optic-based infrastructure has evolved. Known today as the optical carrier (OC) CSI, it followed the template established by the continuing development of the DS and PSTN CSIs. In addition, it further extended the DS infrastructure by using dedicated and channelized bandwidth techniques. Not surprisingly, the former DS series of standards was used as the model for determining how to calculate increases in bandwidth thresholds over fiber-optic cable, how to extend the geographic coverage areas (including areas not serviced by the DS network), and how to finalize the standards for OC bandwidth threshold levels for the transport services to be provided through the OC carrier services infrastructure.

When data network standards for LANs, MANs, and WANs were developed in the mid-to-late 1980s and external transports were needed to interconnect various LAN and MAN sites, both the DS and OC carrier services infrastructures were able to rise to meet this challenge. Beginning in the 1990s, carriers elected to install fiber-optic cable whenever possible to supply the transport demands of their customers. T1 and T3 lines formerly based on copper were now being carved out of much larger bandwidth transports of the optical carrier CSI. Figure 4-4 is a good example of this.

VoIP transports go fiber-optic

In the early 1990s, the fiber-optic-supported ATM (asynchronous transfer mode) transport service evolved. Before VoIP, ATM was the only dedicated network type that integrated data, voice, and video applications on the same network transport. Not long after the inception of ATM, some manufacturers developed an ATM option that could be deployed for a LAN solution. But by the time the design costs were calculated for the infrastructure, the overall cost was higher than any other LAN solution available.

ATM ended up competing with Ethernet, and Ethernet won. ATM was developed on the communications side of the fence and Ethernet was adopted on the data (computer) network side. In the beginning, Ethernet was not as fast as ATM; it ran only on slow local area networks. However, over time, Ethernet protocols were adapted to faster transports, such as T1 and T3. Over these higher-speed transport lines, Ethernet was more economical because the equipment to implement it was already in place on the data network side of the fence. In addition, the widespread adoption of Ethernet meant that the necessary equipment became cheaper and cheaper because of the volume of users. Thus, the need for ATM was simply "passed by" with Ethernet's faster lines and cheaper service.

VoIP runs on Ethernet LANs, and the savings from running voice and video over the same ATM LAN was not enough to offset the startup costs when compared to Ethernet and VoIP. Today, VoIP cost-effectively integrates data, voice, and video on the same network with Ethernet as the LAN side of the network.

Other VoIP transports

Ethernet is a given on the LAN side for any customer implementing VoIP. However, one of the major decision points for any multilocation company is what to use as the transport on the WAN side to connect all those locations. In the 1990s, ATM running within the OC CSI had the competitive edge because VoIP was not yet mature. Today, this has changed. VoIP can run on the LAN side and operate very well with ATM on the WAN side. Or VoIP can run on several other OC transport services without the need for ATM.

Nevertheless, ATM took off as a MAN and WAN solution for some companies and most of the carriers during the 1990s. Today, ATM remains the major transport service used by most network carriers. As a MAN and WAN transport service, ATM was hailed as the superior transport service in terms of quality of service (QoS), speed, and the convergence of data, voice, and video. ATM's quality of service far exceeded the VoIP alternatives of the 1990s.

At the same time, however, ATM costs were high. Early on, it required a minimum of an OC-3 transport at each location. (A single OC-3 transport runs at 155 Mbps and is capable of delivering more than twenty-four hundred DS0 channels.) Because ATM was so expensive, its largest customer base would continue to be the network carriers, who used ATM to build their architectural presence in both of the dedicated CSIs (DS and OS). The carriers used their ATM networks to lease or sell other data, voice, and video transport services to their customers. Many today use their existing OC service networks to carry the emerging traffic from a fast-growing VoIP marketplace.

Any sizable WAN network running ATM service no doubt has made a large investment in the cost of ATM-related equipment and transports. The good news is that you can run VoIP over such an infrastructure, leveraging the sunk costs of an OC carrier network against the revenue coming from carrying VoIP and other types of traffic. All LANs in your company have to be Ethernet, and each LAN needs to be upgraded to support IP telephony.

The hybrid fiber-coaxial CSI

When fiber-optic cable began to be deployed widely, the cable companies started using fiber to build out their infrastructure. But by that time, much of the coaxial cable infrastructure supporting localized connections had already been established. This is why a large share of today's cable customers have coaxial cable coming into their premises from the nearby telephone pole.

The cable companies' network is today known as the hybrid fiber-coaxial (HFC) CSI. It combines the use of coaxial cable with fiber-optic cable. The HFC CSI may one day be all fiber-optic cable. In its present state, it provides not only cable TV services but also cable modem, one of the two popular methods of broadband Internet access. To run VoIP in your home, you need broadband service. If you have cable modem service, you can usually add VoIP transport services with little or no additional expense added to your existing POTS telephone bill.

The HFC CSI began to evolve in the 1980s as strictly a cable-television application. Companies in the business of supplying closed-circuit cable television programming used satellite technology to capture both broadcast television signals from far-off places and local TV signaling, and pipe those TV signals through their cable-based infrastructure to consumers willing to pay for the better quality and channel selection.

Companies in the cable TV business had to bear the cost of building out the HFC infrastructure because there was nothing in place when they first got started. The cable carriers utilized many of the inground conduits and telephone poles already in use by the OC and DS CSI carriers. They also built buildings and facilities for terminating cable services. When the consumer demand for broadband Internet access developed in the early 1990s, the HFC CSI was in a reasonably good place to integrate Ethernet access — and therefore VoIP — within the home using their existing cable television network. Today, broadband Ethernet running over the HFC carrier network has more than twenty million customers and is growing rapidly each month.

The wireless CSI

Wireless telecommunications have been around for more than eight decades. First we had the radio in the 1920s. During World War II, we had the inception of walkie-talkies. These led to the development of cell division multiple access (CDMA), one of the most popular carrier services supporting cell phone networking today. In the 1960s, the first wireless transports connected mobile telephones using radio telemetry, which connected the caller (using radio frequency channels) to the circuit-switched PSTN.

Wireless telephones used radio telemetry until the first cellular network towers began to evolve in the early 1990s. The wireless telephones went through many variations, with each iteration getting smaller, cheaper, faster, and better. Wireless telephones first used analog modulation, then digital and hybrid techniques — and even satellites.

How does VoIP fare with the wireless CSI? The jury is still out, but at this point little can be done with VoIP over cellular networks. Why? Because the cellular network, even though it goes over the wireless CSI, is essentially an extension of the PSTN.

However, there are two exceptions to this. First, a computer could be running a VoIP soft phone application (see Chapter 10), which allows the computer's user to be connected to a VoIP network and conduct voice conversations through the computer connection. The computer's connection to the Internet or to a company's WAN could be established through a cellular data service. (Many cellular telephone companies are now offering high-speed data connections for their users.) In this case, VoIP is being operated through a cellular connection, which means it is going over the wireless CSI.

The second exception is for dual-use telephones, which can access both cellular networks (the wireless CSI) and VoIP over wireless computing networks. These phones are able to place VoIP calls over a wireless data network when one is within range, and over the regular cellular network when one is not.

It is easy to confuse wireless networking with the wireless CSI. They are not the same. Wireless networking is an extension of Ethernet networking, and is discussed in depth in Chapter 8. The wireless CSI is, today, the cellular network used predominantly for voice communications.

Summing up the CSIs

A CSI is like a highway system that lays out all the many roads that enable people to drive to their destinations. Within our highway system, we could characterize some roads as being large or wide, some roads as small or narrow, and some roads as being between these two extremes. Roads may be further broken down by type of surface, that is, asphalt, concrete, gravel, or dirt.

Similarly, we could characterize a CSI as having great amounts of bandwidth capacity or limited bandwidth capacity; as single channel or multichannel; as switched or dedicated; or as circuit-switched or packet-switched. Table 4-1 summarizes the overall state of CSIs.

Table 4-1 Carrier Services Infrastructure Types and VoIP Services

CSI	Inception	Network Type	VoIP Transports	VoIP Service Options
Public switched telephone network (PSTN)	1879	Switched	PRI line DSL using POTS line	VoIP over PRI VoIP over broadband DSL (VoDSL)
Digital service (DS)	1964	Dedicated	DS1 (T1), DS3 (T3)	VoIP over private dedicated network channels
Optical carrier (OC)	1980s	Dedicated	OC3, OC12	VoIP over private dedicated network channels Used to provision other dedicated transports such as DS1, DS3
Hybrid fiber-coaxial (HFC)	1980s	Dedicated	Cable fiber	VoIP over broadband cable modem
Wireless	2003	Switched	Frequency spectrum channels	VoIP soft phone for pocket PC VoIP over WiFi (VoWiFi) VoIP over WiMax (VoWiMax)

VoIP runs best in a dedicated, packet-switched carrier services network. For a company with multiple locations, this means primarily using transports coming out of the DS and OC CSIs. Wireless transports may be used to augment or support the routine need for remote telephony services.

Carrier service companies are constantly adding and upgrading network transport lines and equipment in all five of the CSIs. They also grow by merging with carriers that are more heavily vested in another CSI than they are. This is important to understand if you're running VoIP in a multilocation network. If you have private, dedicated transports, you're not so much concerned with how much of the dedicated line is owned by one or more carrier providers as you are with the underlying requirement that it be dedicated to your VoIP network 24 hours a day, 365 days a year. At the same time, if you can acquire dedicated lines that are owned from point A to point B by one carrier company, chances are that the single owner may be more apt to resolve maintenance problems than a dedicated line owned by several carrier companies.

Just like the highway system, CSIs are not owned by any one carrier because all carriers own a portion of each CSI. What they do not own they must lease from other carriers at wholesale and resell to the customer. Most carrier service companies can lease network transports from all five CSIs.

Carrier companies have countless miles of copper and fiber-optic cable running through underground conduits across the country and the world. In addition, the carriers strategically locate their facilities throughout the country to terminate and switch out all their network transport lines and wireless channels to best support their customer base. Taken in total, all these networks constitute the five CSIs described in this chapter.

How VoIP and the Internet Fit the CSI Picture

To understand how the Internet relates to the five CSIs, it helps to first recognize what the Internet is. No doubt Bell would be captivated by the enormity of what we call the Internet. By definition, it is a network of networks. But just like VoIP can run on any of the five CSIs in varying degrees of quality, the Internet can also be accessed from any of the five CSIs in varying degrees of quality and security.

VoIP over Internet

Yes, the Internet is a network of networks, and the Web is one of its largest applications. But the Internet is also a network that is accessible through all five CSIs. Figure 4-5 provides an illustration of how this access is provided.

Since the Web emerged into the private sector in the early 1990s, the entire Internet has been converted to a tiered infrastructure that predetermines broadly what kind of quality you can expect over your Internet connection. What tier your Internet provider operates at is a major factor that controls your bandwidth throughput and therefore the quality of your VoIP services. Several new terms surfacing in light of VoIP are VSP, for VoIP service provider, as well as VoIP provider and VoIP hosting provider. (Using VoIP over the Internet is covered in more detail in Chapter 9.)

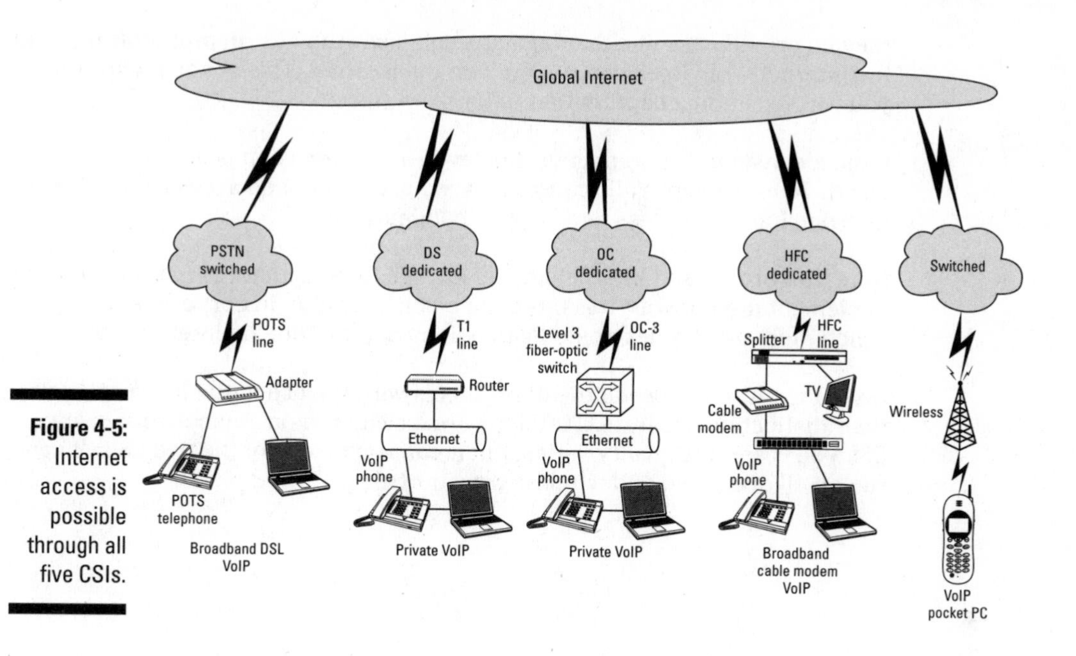

Figure 4-5: Internet access is possible through all five CSIs.

VoIP in the corporate sector

Technically, the Internet is a network (a transport) and VoIP is a network transport service. Both the transport and the transport service are provided in varying degrees of quality on all five CSIs. All five CSIs can provide access to the Internet, just as all five CSIs can support VoIP services.

Keep in mind that if you run VoIP on the Internet, at least one CSI is involved. The Internet is not a CSI: It is a network (global though it may be) that results from all five CSIs providing access to it.

Even though VoIP was developed in 1995, the corporate world did not consider adopting it in a big way until 2001. Issues pertaining to quality and security complicated VoIP's adoption. These issues have been largely addressed, and the corporate sector is rapidly adopting VoIP. The transports available on the DS and OC CSIs are being used most often by companies to set up their VoIP networks to run over private, dedicated network transports.

VoIP in the consumer sector

Another point in common between the Internet and VoIP is that they use the same group of protocols. When some people hear the "IP" portion of "VoIP," they incorrectly assume that because VoIP uses the TCP/IP protocols used on the Internet, VoIP itself can run only on the Internet. This is not accurate, as you will see in the chapters that follow.

From a consumer's perspective, however, such confusion is easy to understand. To implement VoIP, consumers require broadband access to the Internet. Internet access and VoIP go hand-in-hand.

Note that broadband Internet access for VoIP is an artifact of only consumer implementation of VoIP. It is often not a consideration in corporate VoIP, which most often relies on dedicated lines that provide no Internet access.

Do not confuse the use of the IP protocols with the use of the Internet; keep their distinction separate. TCP/IP protocols can run on any network in any CSI. VoIP's service quality varies, but it can run on any of the five CSIs. It can run on all dedicated and wireless transports.

Chapter 5

Getting Switched

In This Chapter

- Supporting VoIP over the PSTN
- Controlling millions of calls
- Understanding quality of service issues

In previous chapters, I discuss the beginnings of the public switched telephone network, or the PSTN. I also outline some history of the PSTN. Perhaps the most significant piece of history was the development and deployment of digital telephony services in the early 1960s. Since then, the PSTN has experienced other significant transport line and service developments, including ISDN and DSL. In the consumer market, these newer transports and services are provided over the existing POTS line, providing greater bandwidth.

Today the PSTN is also known simply as the *switched network*. This chapter provides you with the details you need to understand the other transports in the switched network. Also covered are VoIP services available on the PSTN and how calls are controlled on the PSTN.

Understanding How the PSTN Supports VoIP

No other network in the world can compare to the reliability of the U.S. switched network. (Granted, a handful of disasters have disrupted PSTN services in specific regions, but these are the exceptions, not the rule.) Such reliability, however, comes at a high price: The cost of the switched network, particularly recurring (per-minute) charges, is the highest in the world. Regional toll and international calling using the PSTN are the most highly regulated switched network services. This means high regulatory fees in addition to recurring usage charges. But VoIP greatly reduces and may eliminate these types of charges.

The PSTN-VoIP baseline

The quality of service and high performance of the switched network have rarely been in question in the past fifty years. It is only natural that this quality, which we've come to accept and expect, would be considered a baseline, or standard, that VoIP needs to live up to.

Living up to the quality standards of the PSTN presents a problem for VoIP. Remember that VoIP is unregulated, which means it has no enforceable quality standards. Quite frankly, VoIP can't meet the level of quality set by the PSTN in each and every network design, and therefore VoIP is not for everybody. This will change as VoIP replaces the traditional telephony services and customers demand acceptable quality standards.

As it is now, VoIP runs best when implemented on a private, dedicated network. With this in place, any company can utilize any of the other transports to place and receive telephony calls at low or no cost. (The dedicated network options are covered in detail in Chapter 7.) In this chapter, I clarify the three switched transports (POTS, ISDN, and DSL) that may be used to deliver reasonably good quality VoIP to the consumer market, to smaller companies, and to those in the home seeking to connect through their company's larger corporate network.

The POTS transport

As you already know, POTS is a transport that runs through the circuit-switched PSTN. All transport lines in the PSTN have a circuit-identification number, which is either all numeric or alphanumeric. For example, a POTS telephone number has an area code, a prefix, and a suffix that correspond to the physical circuit and the lines that make up that circuit.

Although POTS does not run VoIP directly, POTS is required for the later digital transport, DSL. Because of the need for a POTS line to have a circuit ID, you must have a POTS line established before you can order broadband DSL.

DSL runs on the same line as your POTS telephone service. This raises an interesting question. If you are looking to get broadband DSL so you can run VoIP, do you need to have the added cost of the POTS service? For now, you do. I expect this will change as competition heats up and POTS carriers continue to lose consumers to the broadband cable carriers. (More about this dilemma in a moment, in "The DSL transport.")

The ISDN transport

Work on developing ISDN began in the 1970s but would not be sold to the bandwidth-hungry customer until the early to mid 1990s. Many said it was too little too late, and the consumer market for ISDN never took off. After the news of the first VoIP telephony call over the Internet spread in 1995, a renewed interest in ISDN emerged for a short while. But by this time our attention was turned to the emerging DSL technology first deployed in 1998.

The eventual ISDN standard provided for two flavors of ISDN: Basic Rate Interface (BRI) and Primary Rate Interface (PRI). The ISDN standard defines the basic unit of bandwidth as a B channel, which provides 64 Kbps of bandwidth. *B* stands for *bearer* channel, which is another name for the channel that carries POTS calls over the PSTN.

In the digital world, all transport lines provide one or more channels, just like your cable television provides different channels to carry various programs. Unlike POTS calls, ISDN calls originate in digital form and travel over the switched network to the destination being called.

BRI

By the time ISDN rolled out to the public, other transports and services had evolved that provided more bandwidth without the complexities and cost factors associated with the BRI flavor of ISDN. Some BRI customers are still out there, but they are usually in the process of converting to DSL, cable modem, or some variation of wireless technology. The monthly recurring charges for BRI transport services are considered exorbitant — and are even higher than POTS recurring charges. As a result, BRI has generally dropped out of the VoIP picture.

PRI

The PRI implementation of ISDN has proven to be a cost-effective transport option for companies with a single location seeking to run IP telephony on their LAN. PRI supports local POTS calls into the PSTN.

PRI is the one ISDN transport that has remained useful for supporting VoIP. It provides a customer with 23 B channels of digital bandwidth. In addition, the carrier configures the PRI to have one 64 Kbps D channel, which is used to manage the line. The industry summarizes the PRI aggregate digital bandwidth as "23 B + 1 D."

In general, a VoIP call is made over a PRI transport as follows:

1. **A caller uses their IP-enabled phone to dial a number.**
2. **The number initiates a packet on the LAN. For inside calls, the packet stays on the LAN. For outside calls, the packet is switched to the PSTN gateway.**

   A *PSTN gateway* is a hardware device whose basic function is to sit between two dissimilar networks and translate the packets that pass through into the format required by the destination side. Many different levels of gateways include other functions, such as routing and network management.

3. **The PSTN gateway may use a PRI transport to connect directly to the PSTN. When the gateway gets the call request, it allocates a B channel on the PRI transport, initiates the call, and passes the call to the PRI transport.**
4. **After the call is on the PRI, it is translated for operation over the PSTN as a circuit-switched call.**
5. **When the call reaches the destination telephone, a circuit is established between the caller and the receiver for the life of the call.**
6. **When the call is complete and either party hangs up, the PRI B channel is returned to the channel pool controlled by the PSTN gateway.**

Figure 5-1 provides an illustration of a network layout that could support this call scenario using a PSTN gateway.

Because PRI is a switched transport, it easily connects to the PSTN, which is also switched. PRIs are compatible with the call control used on the PSTN to manage POTS-related calls. (Call control is discussed at some length later in this chapter.)

In effect, the PRI is capable of handling twenty-three individual telephone calls simultaneously, delivering an aggregate bandwidth capacity of 1.472 Mbps over the PSTN. In addition, the PRI transport can be used for computer data as well as videoconferencing. The PRI transport continues to be employed because it is effective, compatible with the PSTN, and cheap, averaging about $275 to $425 per month. Local recurring telephony charges still apply, as they do with POTS or any off-net VoIP call.

The PRI is not a good fit for every VoIP network environment. But for a single-location LAN running IP telephony (VoIP on the LAN) with separate Internet access, the PRI is an effective solution for gaining access to the PSTN for your network.

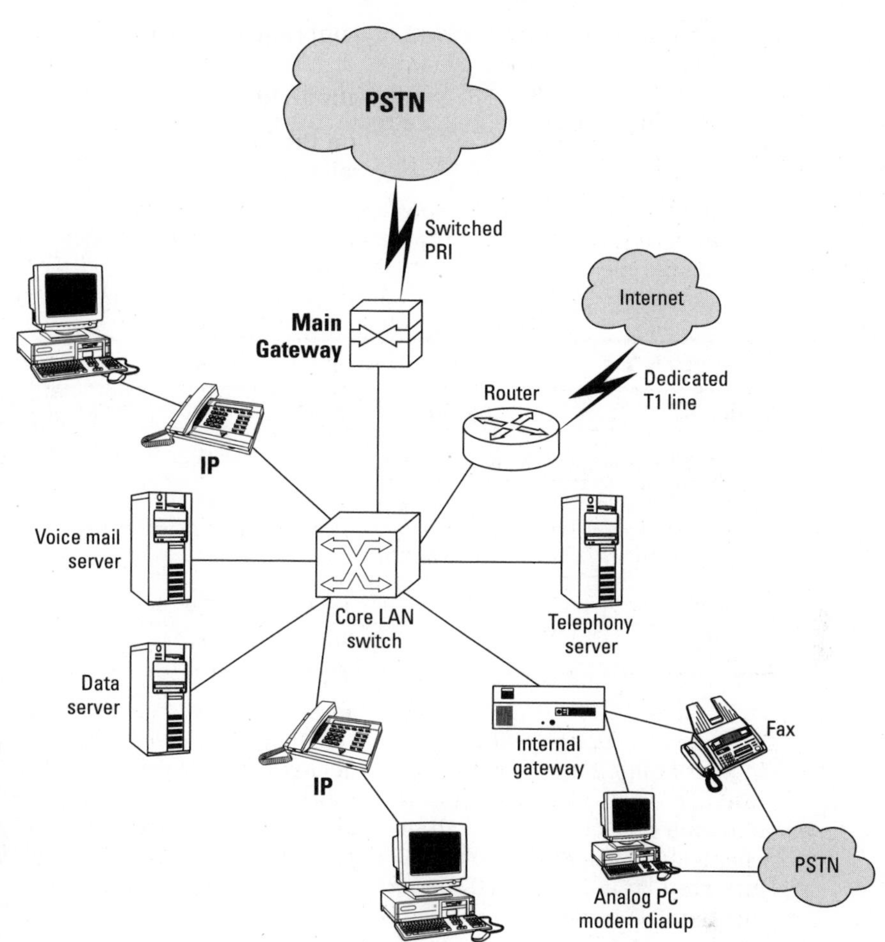

Figure 5-1: Making VoIP calls over a PRI transport.

The DSL transport

DSL stands for *digital subscriber line,* a form of broadband service primarily used by the consumer market and those who telecommute from home. (DSL is also used by some companies for data-only Internet services.) DSL has become one of the most popular transports for running VoIP in the home.

Many criticize the fact that DSL customers must have a POTS line with an assigned phone number. A counter argument is safety — and backup phone service. If you keep your POTS phone plugged into your broadband service, you can call 911 directly. (Consumer VoIP carriers are only now being required to deliver real-time local 911 services, but it will take at least two to three years for the service be uniform and operational.) In addition, if your VoIP provider's network should go down or lose power, you can still make telephone calls using your POTS phone. The main benefit of having VoIP in the home at this time is to eliminate recurring toll charges, not to replace 911 or local calling.

So what can a user expect to pay for getting the benefits of VoIP in their home? Table 5-1 details the typical costs of home VoIP service, when running over a DSL line.

Table 5-1 Costs for VoIP over a DSL Transport

Item	*One-Time Charges ($)*	*Recurring Charges ($)*
POTS line	65	25
DSL service	100 (waived)	50
VoIP adapter	225 (waived)	0
VoIP service plan	35	50
Total	100	125

As you can see, it doesn't take much to financially justify having VoIP service in your home. This is particularly true if you are already using DSL as your Internet access method. Thus, the monthly cost that is strictly VoIP is the one-time service-plan activation charge of $35 — and the monthly recurring charge of $50. If your monthly toll usage charges for intralata, intrastate, interstate or international run higher than this amount (or more than $600 per year), you are money-ahead by getting VoIP.

Controlling Calls

You might wonder how the PSTN manages to control millions of circuit-switched telephone calls each day. Most people know that their telephone connects to the public telephone network, but they don't know what happens to make it work beyond that point.

A critical part of the PSTN infrastructure is making each telephone call successful. To do this, there needs to be some mechanism by which a call is initiated (sometimes referred to as *call setup*), maintained, and terminated (sometimes referred to as *call tear down*). Call setup establishes a channel over which communication can occur, and call tear down releases that channel so it can be used by a different call. These steps — setup, maintenance, and tear down — are collectively referred to as *call control,* or *CC* for short.

Call control does not apply to VoIP on-net calls. When you are operating within the PSTN or are initiating a call on a VoIP network and the destination is on the PSTN, call control becomes relevant. In this section, I discuss various aspects of call control, with particular emphasis on how it affects VoIP.

Signaling system 7 (SS7)

We have come to expect a certain degree of quality with POTS telephony. That quality is ensured by the use of signaling system 7 (SS7), the call-control protocol. SS7 assigns a separate channel in parallel to each PSTN call and provides call control information through this separate channel. That information is responsible for maintaining the call so that the line doesn't "go dead" while you are talking with someone and so you don't get other distractions such as line fade or crosstalk.

SS7 is also responsible for the high-quality accounting services that support the billings and usage involved with every PSTN telephone call. In summary form, customers are able to manage their aggregate monthly telephony usage costs, assess their monthly recurring line access costs, find out what regulatory fees are being applied monthly, and even determine who is and isn't using the telephony system and what they are using it for. SS7 makes all these reports possible.

A good analogy for SS7 is the air traffic control network in the United States. To control the thousands of hourly flights, a parallel network provides real-time information that the air controller staff uses to control the flights in their respective regions. Figure 5-2 illustrates the parallel nature of SS7 in relation to regular phone calls.

Call control and VoIP

Call control on the PSTN is one of the reasons why today's quality of service is so good — and one of the historical impediments to VoIP. The obstacles presented by call control were overcome through improved technology such as PSTN gateways and newer call-control methods that support dedicated packet-switching calling into the PSTN.

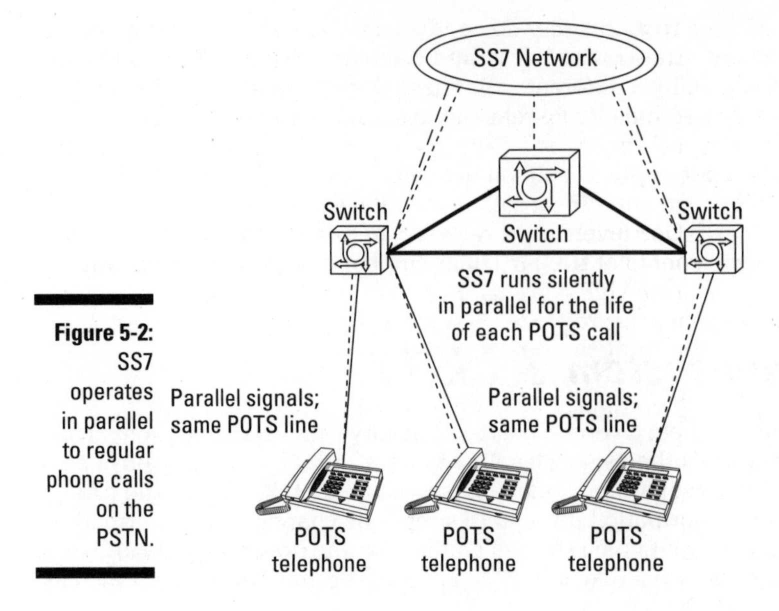

Figure 5-2: SS7 operates in parallel to regular phone calls on the PSTN.

Achieving these newer forms of call control, however, is both complex and costly. For this reason, many companies are opting to design their networks to use their private dedicated transports to support on-net VoIP telephone calls as much as possible. Companies with large multilocation networks that cover the entire country (or even just one or two regions of the country) can design their VoIP networks to route calls destined for other calling areas as far as possible over their private network before going off-net to the PSTN. This type of design optimizes private network use, reduces or eliminates costs incurred with the PSTN, and still provides the QoS benefits of the PSTN.

Only in cases where the company has exorbitant local recurring PSTN charges does it make sense to consider changing how calls into the PSTN are controlled. When a call must go off-net, converting from the packetized VoIP network to the public-switched network is at the very heart of converging the calls.

Delays and errors

When converting to VoIP, another call-control consideration pertains to controlling network errors and delay. In data networks, delay is not a big deal because the network can compensate for it by reassembling packets at the destination. If that fails, a well-designed network can request that the data be retransmitted.

Even though all the error correction and retransmission can be performed at top network speeds, data packet speed transmission requirements are much slower than those required for transmitting VoIP packets. If I don't get an e-mail message for a couple of minutes, this is no major disruption. But if there is a delay in a real-time voice transmission, it messes up the quality of the conversation.

To work at the same QoS levels as POTS, VoIP demands real-time speed that exceeds the requirements of traditional data networking. That is why you can't adequately operate VoIP on networks that run at less than broadband speed. Unlike data packets, VoIP does not retransmit packets. What you hear (and see, when transmitting video) is what you get.

A big part of the need for minimal broadband speeds pertains to the underlying requirements for network design. As covered in Chapter 1, the computer network uses the TCP/UDP/IP networking model and its related protocols to support both data packets and VoIP packets that travel the network. Keep in mind that any network connection can and usually is several network connections strung together to ultimately provide the end-to-end connection. As a result, the design of the network used to support VoIP is critical to the QoS that you have with VoIP telephone calls.

Quality and VoIP

When we talk about various forms of telephone quality of service (QoS), it is understood in the industry that *toll quality* means the highest form of telephony service quality controlled by SS7. One of the largest problems with early VoIP was QoS. Connections could be made over the PSTN and into the Internet, but what callers heard was far from what we've become accustomed to with the toll quality of POTS. Since then, we've learned that VoIP QoS comes down to controlling specific errors that are commonplace on VoIP networks. Today, all of those errors can be controlled through network design.

VoIP telephony service using DSL is said to be near toll quality. With a private, dedicated approach to VoIP, your company can achieve toll quality (or better) VoIP service.

VoIP hardware and software is improving to the point where error rates will soon be a thing of the past on all network types. Even so, it is important to understand the three factors that can affect QoS on a VoIP call:

- Network delay
- Poor compression
- Signal attenuation

Network delay

In 1995, the delay for a VoIP call using the Internet ranged from 400 to 4000 milliseconds (ms), or 4/10 second to 4 seconds. This delay was quite noticeable in a conversation and was typical when using VoIP over POTS lines. Fortunately, since 1995, network options have improved, which has reduced delay and thereby improved quality.

Delay continues to be a factor in VoIP network design and management. It is something that network professionals watch for continuously. Actual delay depends on the type of network access, the overall distance between the caller and the receiver, the total number of users on the network, the network type involved in the connection, and even the equipment used.

Leading IP telephone manufacturers and VoIP carriers use a benchmark of 150 ms as the maximum acceptable delay. That delay metric, by the way, was established long ago as the delay benchmark for POTS.

Closely related to delay is a factor known as *jitter.* Whenever VoIP packets are received outside the expected window of time (delayed), jitter occurs. Jitter may be caused by timing delays introduced by equipment or transport failures, increased network traffic, or changes in the configuration of the network. A *jitter buffer* can be used to store packets as they arrive at the receiving end; the packets are then distributed to the destination VoIP telephone in the correct order. In this way, delayed VoIP packets don't disrupt the conversation.

Poor compression

VoIP converts the sound of your voice into packets of data, sends them across the network being used as the transport, and then reconstructs them into sound at the receiver's end of the network. The sound information, often called the *payload,* is put together with overhead information identifying where the packet should go to create the final packet transmitted over the network.

VoIP often uses compression techniques to reduce the total size of each packet. The benefit of compression is that it increases speed and optimizes available bandwidth. As a result, there is less delay and higher QoS.

VoIP is not regulated and is still rather new, so compression techniques used by consumer-market VoIP carriers are not standardized and are still being developed.

VoIP carrier service providers can be differentiated in terms of service quality relating to compression and network speeds. How much of a commitment will the carrier make to ensure a high quality of VoIP service? You want to read in your service level agreement (SLA) how much speed and bandwidth you can expect. Reputable carriers usually list 256 Kbps as the minimum speed and as high as 1.536 Mbps as the maximum speed. If they state that it is "best effort" within this range, that is not so bad. If they state nothing or give no range, move on to a different carrier.

Signal attenuation

Signal attenuation is the degradation of a signal over distance and time. Did you ever drive out to the country or into the mountains with a radio on? As you get farther and farther from the source of the signal, the radio fades out or you hear more static.

This is similar to signal attenuation over the VoIP network. VoIP packets are represented on many transports as a change in voltage, and that voltage can degrade over time and distance. By the time the packets arrive at their destination, the signal inside the packets is no longer in its original form. If attenuation is very poor, the person you are calling may not even be able to distinguish who you are or what you are saying.

Attenuation can be reduced or eliminated by improving the speed at which packets are delivered on the network. One way to fight attenuation is with compression, discussed in the preceding section. The better the compression, the better the attenuation.

Chapter 6

Going Broadband

In This Chapter

- Transmitting over broadband
- Setting up VoIP over cable
- Setting up VoIP over DSL

In the beginning, the only public access to the Internet was through slow, dialup modems, which had a typical maximum speed of 56 Kbps. During the 1990s, people clamored for higher-bandwidth alternatives. Much of the demand centered on access to the Internet for applications such as e-mail and e-commerce. Companies wanted more bandwidth also to make use of telecommuting applications that remote employees could access through corporate networks.

By 1998, two high-bandwidth options, which had been in development for several years, were finally rolling out to the public. These two types are now known as cable modem (CM) and digital subscriber line (DSL). Because both options increased the available bandwidth when compared to a traditional modem connection, they were labeled as *broadband.* Actual throughput varied, but the minimum speed for both broadband types was 256 Kbps, and the maximum speed was 1.536 Mbps.

At the same time that the media coined the term *broadband,* the term *narrowband* began to be used to characterize older, slower connections on POTS lines. By 2000, the two broadband options had made their mark, and narrowband was well on its way into the bandwidth history books.

Broadband access — either cable modem or DSL — provides an easy way to implement VoIP. From a network perspective, you need only a VoIP adapter box and a service plan to make good-quality VoIP calls. The VoIP adapter box is connected to your broadband service device (either a broadband modem or router). The adapter box provides ports for connecting your VoIP phone or your POTS phone (or both) as well as your computer. In addition, many consumers add a wireless router to the adapter box to gain more ports and wireless (WiFi) connectivity. Each of these devices must plug into a power outlet. I suggest you get a power strip with surge protection.

Broadband Transmission Methods

Two types of transmission are used with both cable modem and DSL broadband services: asymmetric and symmetric. These transmission types are necessary to optimize and share bandwidth, which is more of a requirement in highly populated areas, where the lines can get congested. These two types are distinguished by the amount of bandwidth available to the end-user and how the actual transmission takes place over the transport line.

Asymmetric

Asymmetric transmission is the typical option used for consumer broadband. *Asymmetric transmission* means that the data rate is different depending on the direction of the data. For example, a typical consumer broadband setup may have an upstream of 384 Kbps and a downstream of 1.536 Mbps. The difference in transmission speeds is based on the concept that end-users are normally consumers of information, not providers of information.

VoIP requires both upstream and downstream data transfers. Fortunately, VoIP requires only a 64 Kbps channel to deliver high-quality services, so it can operate over regular asymmetrical lines.

Symmetric

As you might have guessed, *symmetric transmission* means that the upstream data rate is the same as the downstream data rate. Symmetric transmission normally comes with a higher price tag that varies according to the bandwidth desired.

Even though symmetric transmission is more expensive than asymmetric, such a connection may be necessary, particularly if you are running your own Web server or e-mail server over the connection. In that case, incoming traffic is generally low, but outbound traffic — delivering content to users — can be critical.

As with asymmetric transmission, VoIP requires a minimum of 64 Kbps to operate. If your symmetric connection is faster than this minimum, you'll do fine operating VoIP over the link.

Broadband by the numbers

The Federal Communications Commission reported at the end of the first half of 2004 that the number of broadband transport lines in the United States had reached 32.5 million. The report indicated there were 11.4 million DSL users, 18.6 million cable modem users, and more than 2 million other users (satellite and miscellaneous connections).

The market for broadband lines is still wide open. I have no doubt that the maturation of VoIP over broadband Internet connections is persuading consumers and small businesses to adopt VoIP. I can see the day when satellite providers will market "broadband," and many will think that it is actually ultrahigh-speed Internet access. Apparently any network transport service, hardwired or wired, that accesses the Internet at a speed faster than a POTS line is a candidate for the broadband category.

VoIP with Your Cable Modem

Cable modems provide Internet connectivity through the HFC CSI (described in Chapter 4). The HFC CSI was originally designed to deliver television signals to the residential market. As the need for residential Internet connections exploded, the cable industry optimized its network to provide high-speed Internet access. VoIP then became a possibility for this market.

VoIP shares something with your TV

Today the HFC CSI is used not only for cable television, but also for traditional POTS telephony and data services such as Internet access. These services are sold and billed separately on a monthly basis. Based on the cable company, customers may be able to choose which services they use. Traditionally, companies have required that users have at least basic cable service, but such a requirement is being dropped by some companies. In some areas, it is possible to get just telephone service or just Internet service over the coax cable traditionally used to deliver the television signal.

You may wonder how VoIP can run on the same cable that provides cable television. Just as the cable can deliver hundreds of different television channels, it also supports the assignment of channels for Internet data services. For POTS telephony service, the cable companies can pick off the POTS signal and divert it into the PSTN.

Because VoIP transports voice signals inside IP packets, the cable infrastructure doesn't need to distinguish regular data packets from those packets transporting VoIP. After the customer is connected to the Internet, the VoIP provider takes care of sending and receiving the VoIP packets. The VoIP provider may or may not be the cable company.

Adding VoIP

The cable customer is free to enter into a VoIP service agreement with any Internet-based VoIP provider. As of this writing, VoIP service plans range from $20 to $50 per month, depending on the number of minutes desired per month. Some VoIP providers have unlimited minute plans. The VoIP carrier provides the VoIP adapter box at no cost provided you sign a term deal (such as two years), or you may be required to purchase or lease a broadband VoIP adapter box. Make sure you read the details in the service agreement.

Table 6-1 illustrates the typical costs associated with running VoIP over a cable modem connection. Remember that the amounts shown in the table vary depending on where you live and what services you choose.

Table 6-1 Costs for VoIP Using a Cable Modem

Item	*One-Time Charges ($)*	*Recurring Charges ($)*
Cable TV line	35	40
Internet access	0	50
Cable modem	0	3
VoIP service	30	30
VoIP adapter	0	0
POTS services	0	15
VoIP toll service	0	0
Totals	65	138

Setting up VoIP on a cable modem

After your cable modem is in place and functioning properly, adding VoIP is relatively easy. Figure 6-1 illustrates the typical method for connecting VoIP through a cable modem.

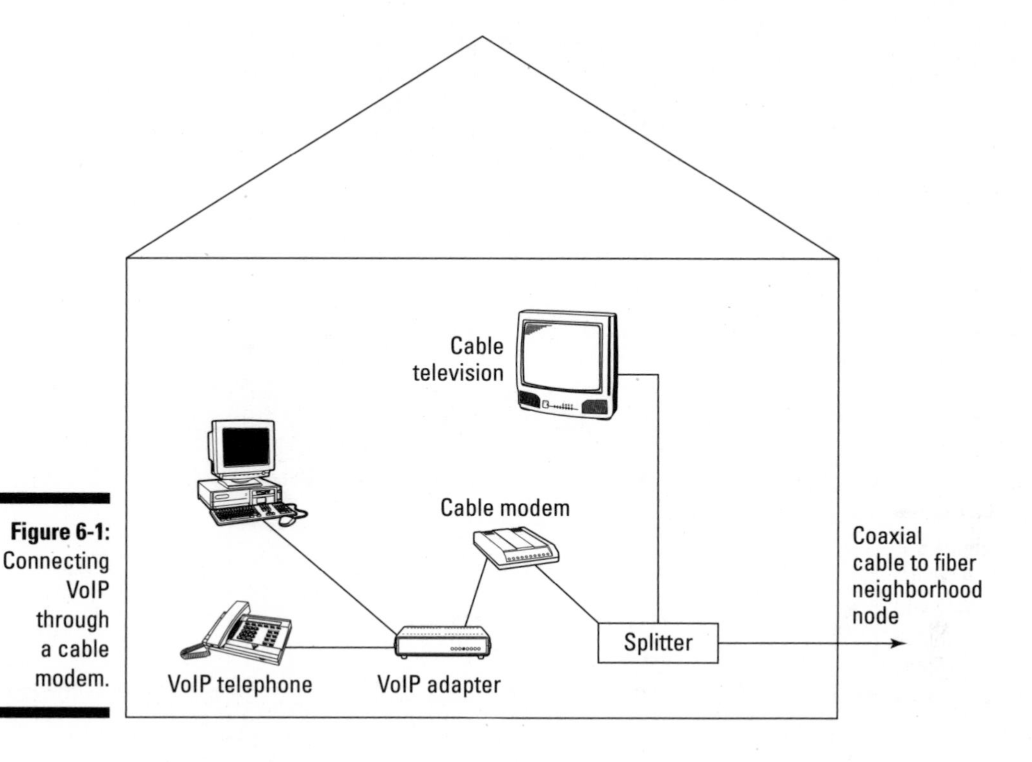

Figure 6-1: Connecting VoIP through a cable modem.

Following is the procedure for setting up VoIP service over your cable modem:

1. **Connect the VoIP adapter box to the cable modem.**
2. **Connect your computer to the VoIP adapter box.**
3. **Install the software provided by the VoIP provider.**
4. **Configure the telephone number and other parameters as directed by the VoIP provider.**
5. **Connect a microphone and speakers or plugs into the VoIP headset or handset.**
6. **Use the IP soft phone directly from your computer and begin to place VoIP telephone calls.**

An IP soft phone is software that provides a way to use your computer as a telephone. The software displays an easy-to-use dialing pad on the screen, along with controls for other common telephone functions, as shown in Figure 6-2. Some versions of soft phones provide an interface to a common directory, such as one maintained in Microsoft Outlook.

Figure 6-2: A typical VoIP soft phone dial pad.

Possible cable modem problems

It's not uncommon to get a thirty-day trial period during which you can, for any reason, cancel your cable modem agreement or your VoIP over cable modem services agreement and stop using VoIP. If you have no trial period, the next best thing is a month-to-month plan. Note that you will need to pay an activation fee. If you can't get a trial period, record any problems you have with the service during the first month.

If the service doesn't work for you, you can terminate the trial or make the case that your activation cost should be refunded. In this section, I describe some things to be on the lookout for during your first month of VoIP service.

Network contention

The HFC CSI is *contention-based,* which means that it offers limited resources that are allocated to customers based on demand. It essentially assigns channels to customers as they come online. If you're one of only a few customers in your neighborhood, you will enjoy very good bandwidth and fast data transfer speeds. However, if many users in your neighborhood have cable modems, and you are among those who come online later in the day, you can expect delays due to greater contention for the available resources. The amount of delay depends on how the cable company built the network in your neighborhood and how they manage their network.

The HFC CSI continues to rely on coaxial cable to transport signals. However, research is being conducted on the emerging FTTH (fiber to the home) standard, which would use ultrahigh bandwidth (100-Mbps) fiber-optic cable to the home. FTTH uses a star topology (see Figure 6-3), which has the advantage of better fault isolation, thereby minimizing or eliminating contention and the resulting throughput problems.

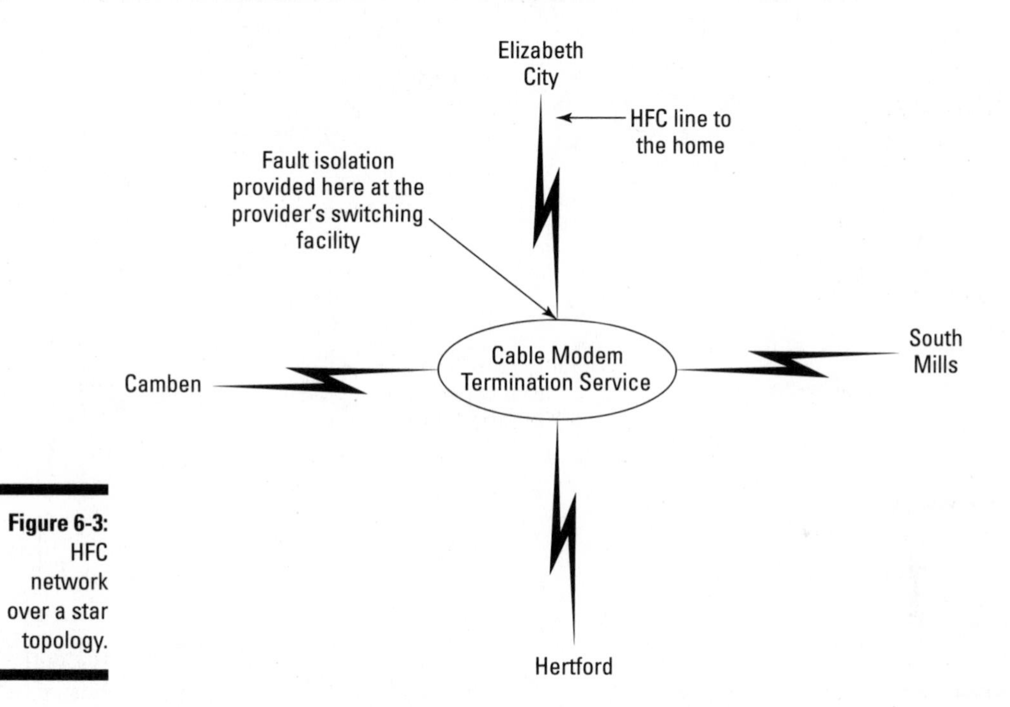

Figure 6-3: HFC network over a star topology.

Potential network failure

Because the older parts of the HFC CSI uses a ring topology (see Figure 6-4), if one part of the network ring is down or not working properly, the entire ring is affected. Moreover, if there is a power failure in the neighborhood, everyone on the network is down. This means any services going through the cable network — including the telephone and Internet — are unavailable.

Mislabeling wireless

In just six years, broadband has become the most popular method to support consumer VoIP. You can even use wireless connectivity (WiFi or WiMax) to implement VoIP. But you should not confuse this wireless connectivity with broadband. They are complementary, but they are not the same. The media, more than any other entity, has often incorrectly called the newer wireless technologies *broadband.* Both WiFi and WiMax technologies enable users to connect wirelessly to the Internet or to any network supporting wireless. The fact that they provide wireless access to the Internet does not change the fact that WiFi and WiMax both implement Ethernet connections, the same as wired networks do.

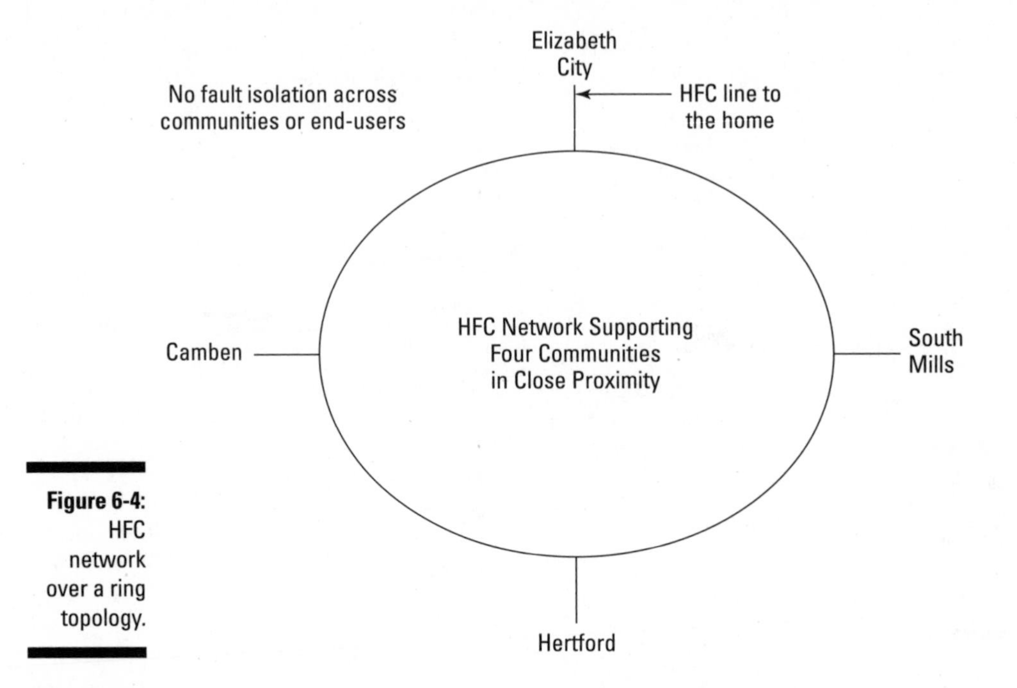

Figure 6-4: HFC network over a ring topology.

VoIP Through Your DSL Connection

DSL, described in Chapter 5, is a high-speed data connection method that uses the ordinary telephone lines that come into your home or business. Utilizing a star topology (see Figure 6-5), DSL provides two benefits important to VoIP: excellent fault isolation and better throughput due to less resource contention.

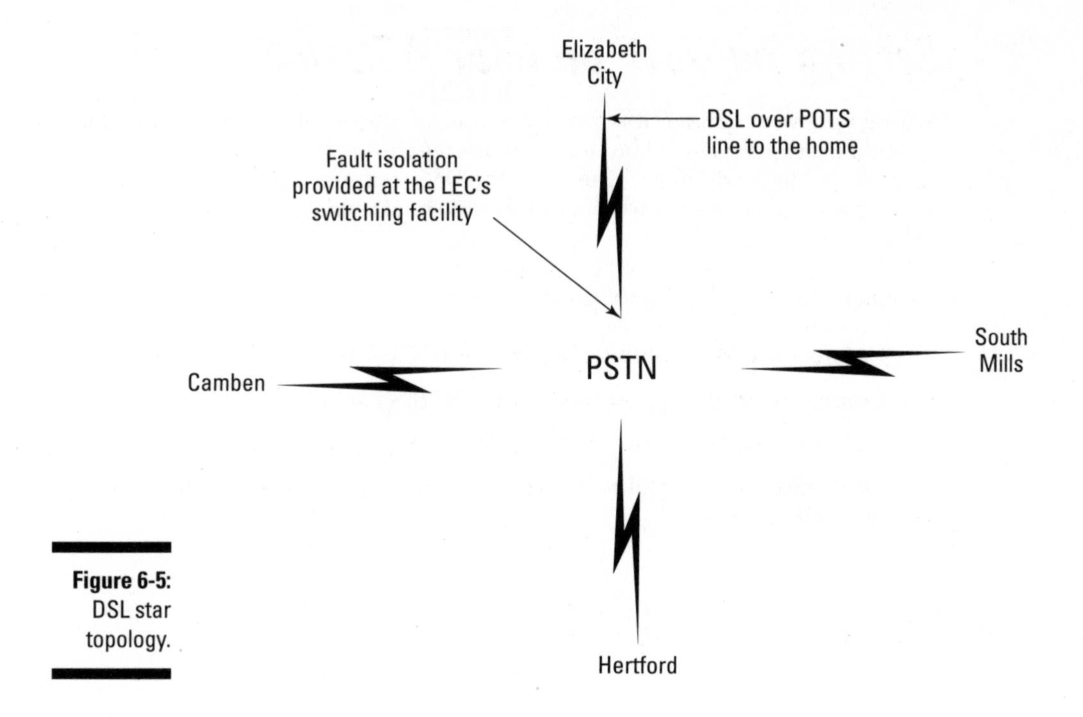

Figure 6-5: DSL star topology.

POTS plus!

DSL utilizes the PSTN CSI and is therefore a switched transport service. All customers must have at least basic POTS telephone service to use DSL. After you are running both POTS for telephone service and DSL for data, you can enter into an agreement with any VoIP provider to obtain VoIP services.

As with using VoIP over a cable modem connection, VoIP over DSL plans average about $50 per month for unlimited calling. (Refer to Table 6-1.) The VoIP provider either gives you a VoIP adapter box or requires you to purchase one. You can keep the POTS line service while running VoIP through the adapter box, essentially making two calls at the same time on a single phone line.

If the VoIP provider you select doesn't have an unlimited plan, seeks to limit your minutes, or charges you by the minute, I suggest you consider a different provider.

Setting up VoIP on your DSL line

Setting up a VoIP connection through your DSL line is similar to connecting through a cable modem. The primary differences are that you use a different adapter and that adapter connects to the DSL router or modem rather than to a cable modem. Figure 6-6 illustrates a normal VoIP connection through a DSL line.

To make your connection, follow these steps:

1. **Connect the VoIP adapter box to the DSL router or modem.**
2. **Connect your computer to the VoIP adapter box.**
3. **Install the software provided by the VoIP provider.**
4. **Configure the telephone number and other parameters as directed by the VoIP provider.**

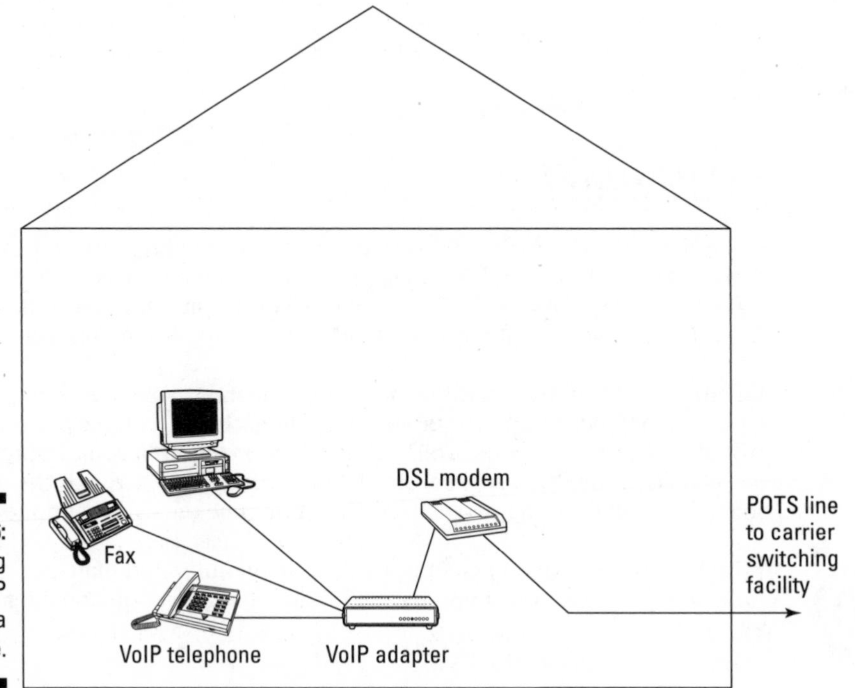

Figure 6-6: Connecting VoIP through a DSL line.

5. **Connect a microphone and speakers or plugs to a VoIP headset or handset.**
6. **Use the IP soft phone directly from your computer and begin to place VoIP telephone calls.**

DSL does not require you to abandon your POTS telephone nor its services. Using an inexpensive splitter that plugs into the VoIP adapter box, you can continue to enjoy POTS telephony as well as VoIP. Figure 6-2 shows a fax line coming out of the VoIP adapter, but this could just as easily be a regular telephone.

Potential DSL problems

The technology used by DSL requires that the distance between your computer and the nearest carrier facility be no greater than 18,000 feet (about 3.5 miles). Distance is a factor with DSL in determining overall data speed: The shorter the distance, the better the throughput. Thus, the closer you are to the carrier facility, the better the quality of your VoIP experience. If you are near the distance limit, you may not be able to get speeds that provide satisfactory quality.

DSL does not provide enough bandwidth to support VoIP traffic for a multilocation company. Therefore, DSL is not acceptable as a transport for a multisite corporate network. However, DSL is more than adequate for consumer and home-business VoIP.

VoIP over POTS

DSL runs over the same PSTN CSI that carries your POTS service. To accomplish this feat, the DSL signal is modulated at a higher frequency than the frequency used for regular voice. Analog POTS lines can't support VoIP for reasons already outlined in Chapter 4, but DSL technology — piggybacked on the POTS line — can support it. DSL uses multiplexing equipment that amplifies, regenerates, and reconstructs the VoIP signals so that the packets can travel to and from the Internet in an acceptable manner.

Figure 6-7 illustrates how your VoIP packets traverse the network through your DSL connection, the PSTN, and the Internet.

A DSL line is like having two channels that are each operated using different networking techniques, both going over the same physical line. One channel is your analog POTS line, and the other is the digital DSL line.

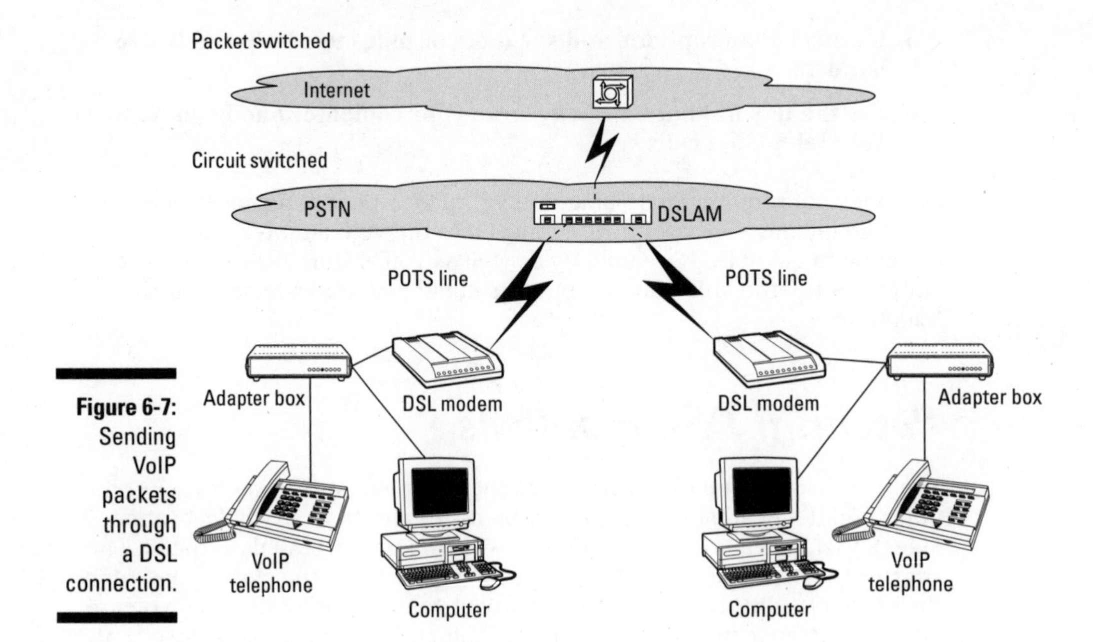

Figure 6-7: Sending VoIP packets through a DSL connection.

Chapter 7

We're Dedicated

In This Chapter

- Understanding dedicated transports
- Transporting over digital service and optical carrier networks
- Converging networks
- Managing your bandwidth
- Keeping a circuit-switched line

As the old axiom states, you get what you pay for. This statement certainly applies to network bandwidth. When you examine the various types of network transports available (see Chapter 4), it quickly becomes apparent that the highest QoS attainable with VoIP is with networks that employ dedicated transports. This chapter examines how dedicated transports can help provide the best VoIP experience.

Basics of Dedicated Transports

Transports are the physical lines installed at the company or consumer premises to provide all sorts of network access. Many folks think of T1 or T3 transport lines when they hear the phrase "dedicated transport," and with good reason. T1 and T3 lines are the most popular dedicated network transports in the country.

VoIP is making dedicated transports even more popular. If you're going to run VoIP on your company's network, dedicated transports give you the best quality VoIP. Dedicated transports also allow you to connect all the data applications used by your company at all your locations.

When transport lines are dedicated, the bandwidth they provide is available fully to the customer at all times. Unlike switched transport lines, which are shared by the public, dedicated lines are 100 percent committed to one specific customer's private network. For this reason, dedicated networks are often called *private networks* or *private line networks.*

Bandwidth and speed

Bandwidth and speed are the darling twins of data networks. When you compare speeds between dedicated and switched networks, you'll find that a switched network transport generally provides far less throughput. For example, a B channel on a switched PRI transport (see Chapter 5) does not provide the same throughput as a DS0 channel on a dedicated T1 transport. (DS0 and T1 are both described in gory detail later in this chapter.)

Throughput is the total amount of data that can be passed over a transport line in a given amount of time. Throughput is directly related to bandwidth and is often used synonymously with data speed.

Two factors affect both bandwidth and speed when it comes to dedicated lines: routing and exclusivity. Routing on a dedicated line is directly between two points, passing through few routers and switches. Data passing through a switched network, such as the PSTN, will go through many routers and switches. The more switching points involved, the less throughput because each switching point adds overhead to track the data.

Exclusivity refers to the fact that a dedicated transport permits only a single customer's data on the line. In a switched network, data is aggregated and shared with others, reducing the bandwidth available to any single customer. Aggregation also involves resource contention, which can increase delay and signal degradation.

Figure 7-1 contrasts the difference between the routes followed by a VoIP call over a switched network and a dedicated transport.

Costs of dedication

Dedicated lines cost more than a regular broadband connection. For companies with multiple locations or heavy data needs, this cost is easily justified by the associated increased bandwidth, availability, and VoIP cost savings.

Dedicated transports, although more expensive than their switched counterparts, continue to come down in price. For example, in Pittsburgh, within a distance of less than 15 miles between two endpoints, network access through a T1 line ranges from $450 to $625 per month with no other usage-based recurring charges. The final cost depends on which carrier installs and leases the line and how much they mark up the price. Switched transports are just the opposite: You pay less for the line itself, but then you must pay the carrier for recurring usage charges according to the tier levels of regulated per-minute charges. A switched PRI ranges from $250 to $325 per month.

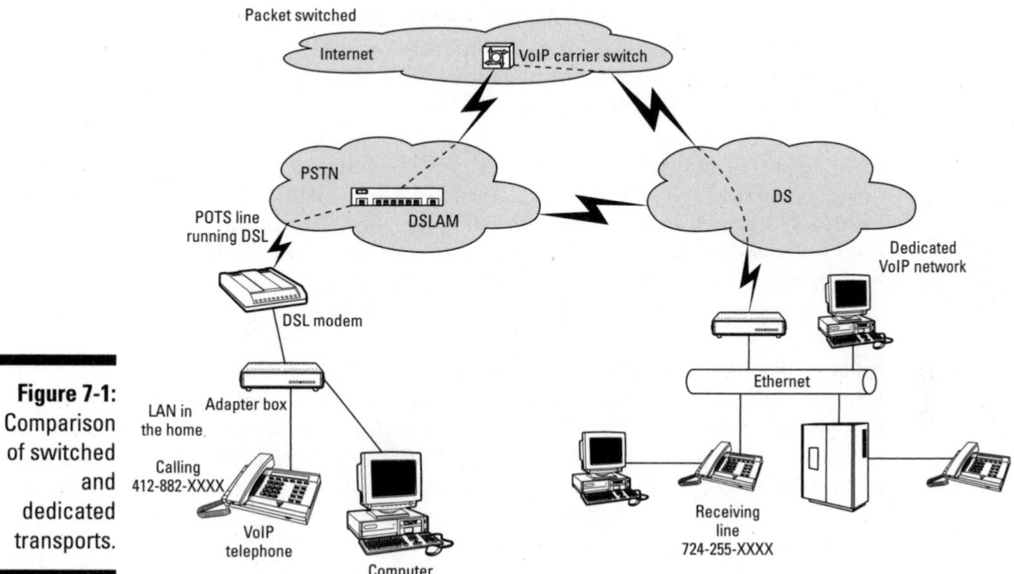

Figure 7-1: Comparison of switched and dedicated transports.

Just four years ago, before VoIP took off, a T1 line cost from $575 to as much as $925 per month for the same distance. In the past, some carriers would discount more if you also leased other services to run over the T1 line. For example, if a customer got Internet access service over a T1 line, the combined cost for this T1 would range from $825 to $1200 per month for similar distances. For a PRI, the cost ranged from $350 to $475, not including recurring usage charges.

One of the reasons for higher costs with dedicated transports is that they run on more expensive cabling systems than do broadband connections. Dedicated transports routinely use ultra-high-speed fiber-optic cable, which is more expensive than copper lines to install and maintain. The extra cost of a dedicated transport makes it more suited for corporate and multilocation company networks seeking to run VoIP because it provides high QoS.

Carriers charge a monthly access fee for the dedicated line that connects your location to the carrier's facilities. This line is called the *local loop,* or *the last mile.* The government regulates what the carrier can charge you for dedicated transports. In addition, you are assessed about 7 percent of the total monthly recurring charge for regulatory fees and taxes. These are the only recurring costs you have for the dedicated transport itself. Basically, these are the costs you pay to obtain access to the local loop.

The local loop is one of the biggest costs in any dedicated network. However, if your company runs VoIP, it can eliminate or greatly reduce costs by putting most of their traffic on-net. In many cases, this savings can pay for the local loop access costs for all locations with money left over.

As an example of how much can be saved, I'll tell you about a large network I designed that had 384 locations nationwide. The client's total recurring charges for the 384 local loops came to $268,000 per month, and their monthly usage charges were $4,200,000 per month. If they converted to VoIP, they would still have the local loop charges, but they could have saved 35 to 60 percent of their usage charges — a whopping savings of $1,470,000 to $2,520,000!

As you can see, the higher costs of a dedicated transport are easily offset by lower or eliminated usage charges. Because a private, dedicated network can be used to carry all of a company's internal telephone calls, the company has no monthly usage charges for internal VoIP telephone calls. On-net telephone calling using VoIP reduces or eliminates traditional telephone usage billings. For calls that must travel off your VoIP network, your company will still have some traditional carrier service bills.

The wireless last mile

The last mile is a hot area for deploying WiMax wireless transport services. WiMax is capable of supporting dedicated access bandwidth speeds at a range of up to 30 miles. It is too early to tell how WiMax will figure into the grand scheme of local loop access. It is also unlikely that first-generation WiMax could support hundreds of locations and thousands of callers at the same time. (The first generation of any new technology seldom does all we want it to.) But second- and third-generation implementations may help drop local loop costs significantly and lead to wide acceptance. In a WiMax world, the carrier companies will have no cable to install between their facilities and the customer premises, so lower labor costs (no installation to speak of) will translate into substantial cost savings.

Real-time network management

The quality of any network is affected by errors and contention. These are normal factors in any VoIP network. VoIP network managers watch the levels of network contention on a daily and even hourly basis. If a manager sees a spike in utilization, it is a sign that something has changed in the normal operation of the network. Perhaps several users are doing massive downloads, or the VoIP telephony calling has increased, or a new server is being installed on the VoIP network. Dedicated transports enable real-time utilization information and management. Such a level of network management is not possible in a circuit-switched POTS-PSTN network.

High-quality VoIP calls

With a dedicated transport, your network can support massive volumes of on-net VoIP calls. Huge networks with multiple locations and hundreds or thousands of callers are best supported with dedicated transports. Dedicated transports enable a QoS that meets or exceeds the quality found in traditional circuit-switched PSTN telephone calls.

Types of Dedication

For the corporate sector, dedicated network transports come in two major flavors: the digital service (or digital signal) CSI and the optical carrier CSI. Chapter 4 covered quite a bit about these CSIs, but you need to know more to make an educated decision about which transport is right for your needs.

The DS transports

The original digital service (DS) series of standards had five levels of dedicated lines. Each DS standard provided a set number of 64 Kbps channels, or DS0s. For example, a DS1 (T1) transport includes 24 channels, and a DS3 (T3) includes 672 channels.

In the old days when DS transports were new and costly, the DS0 was leased as a single 64 Kbps transport line. Today, hardly anyone leases a DS0 transport. If you're implementing VoIP on your company's network, you won't want to consider using just a single DS0 channel. Not only is it not enough bandwidth, but it isn't cost effective.

T1 line

The *T* in the T1 version of transport represents *terrestrial,* or over land. The tariffs controlled by the government for setting the pricing of DS transports are based on the total terrestrial mileage between point A and point B.

The T1 transport continues to be the most popular transport on the market, and prices continue to drop. A big reason why the T1 is so popular is that it permits network configuration folks to divide the total available bandwidth into smaller individual channels. This makes the T1 particularly suitable for VoIP networks that run computer data, telephony voice, and even videoconferencing over the same network transport. T1, however, does not provide adequate bandwidth for large multilocation networks with hundreds of users in each location.

If a company has fewer than three hundred employees per location, one option is to multiplex, or bond, multiple T1 lines to act as one big T. For example, if you multiplex (MUX) six T1s, your effective bandwidth would increase to about 10 Mbps. The costs in this scenario should be considered carefully. The T1 lines would cost about $3000 per month. Additional one-time costs include the six DS1 interfaces required to terminate the T1 lines at your location. In addition, there is the cost of the MUX equipment.

T3 line

If you need more bandwidth than what you can obtain from a T1 or a group of T1s, consider the T3 transport. The T3 transport provides a total aggregate bandwidth of 45 Mbps. This breaks down to about 672 DS0 channels. A growing company can also consider upgrading to T3 transports or some mix of T1 and T3 lines. (The latter is more commonplace for larger companies, which use T1 lines for smaller locations and T3 for the larger locations.)

I had the opportunity to design a crosstown network that required a T3. The client was a small college in the middle of downtown Pittsburgh. They wanted to connect their two high-rise dormitory apartment buildings, and they also wanted Internet access on the network. The total distance between the college and the apartment buildings was three miles. The monthly charge for the T3 line was $19,000. A T1 line for the same distance was $450. In addition, the one-time cost of terminating the T3 was $5000, much more than the termination costs of a T1 line. For terminating a T1 line, even if the client had to buy a new router with DS1 interface, the cost range would be $800 to $3000.

The OC transports

The advent of fiber-optic cabling in the 1980s changed the way that DS lines were installed. By the 1990s, most dedicated transport lines were going in as fiber-optic cables or being implemented through existing fiber lines. The terminating equipment would then be programmed to deliver the equivalent of however many DS0 channels were needed (1, 24, or 672).

Fiber-optic transports are defined according to the OC (optical carrier) CSI standards. The four most common transports are OC-3 (155 Mbps), OC-12 (622 Mbps), OC-48 (2.5 Gbps), and OC-192 (10 Gbps). These types of dedicated transports are used by only the largest corporations and the carriers themselves. The most popular OC standard is OC-3.

Not surprisingly, most small and medium-sized companies rarely have the need for even an OC-3. But carrier companies put in a minimum of an OC-3 whenever they install carrier services for multitenant buildings and the largest customers. Their rationale is to put in enough fiber bandwidth to cover any future customer needs. When those needs arise, the carrier then simply programs their equipment to deliver the necessary bandwidth.

Recent advances in optical multiplexing have taken the OC CSI to new thresholds. For example, Dense Wave Division Multiplexing (DWDM) can achieve bandwidth thresholds in the terabit (trillions of bits per second) range. Such technologies are most often used by carriers to transport data between metropolitan areas across their network backbones. (A *backbone* is a major communication line used to carry the majority of a carrier's traffic.)

Converging Dedicated and Switched Networks

Network convergence is a tricky term that means to bring together or to integrate two or more diverse network types. Everyone considering VoIP needs to be concerned about converging their digital network (the heart of VoIP) and their PSTN (voice) networks. Convergence takes two or more varying modes of carrying telephone calls and enables them to operate together.

VoIP brings together the switched and dedicated networks like never before. Other attempts to converge the two have been incomplete solutions. For example, using a dedicated T1 line to carry POTS telephone calls to the PSTN is not the same as a packetized VoIP telephone call. And POTS-equivalent calls over a dedicated T1 line may still be billed as a regular POTS call. True convergence of the two carrier networks means to merge the packets of a VoIP packet-switched phone call with the circuit-switched signaling of the PSTN.

Figure 7-2 illustrates the convergence of the PSTN with the DS carrier network. The convergence is made possible through the use of a network-to-network interface (NNI). NNIs are typically very expensive and are used by carriers and only the largest customer networks, which have the volume to justify the high monthly cost.

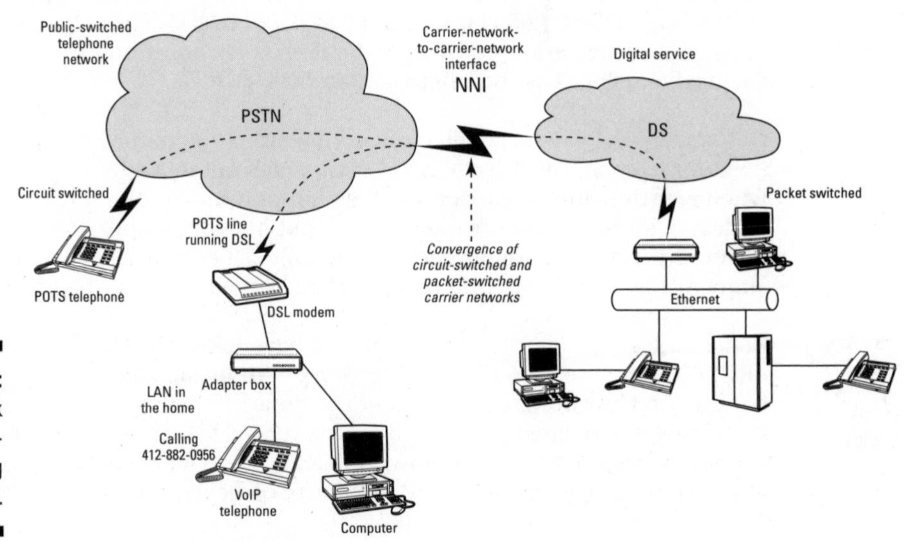

Figure 7-2: Network convergence using an NNI.

A less expensive option does exist for connecting your VoIP network to the PSTN. Figure 7-3 illustrates how a company can converge their network onto the PSTN using a far more cost-effective PRI transport option.

In this example, on-net VoIP calls are carried over the company's dedicated network. Outgoing local calls travel on-net as VoIP packets and then travel to the PSTN through the gateway and its attached PRI transport line. Inbound calls from the PSTN travel as circuit-switched calls. The gateway translates inbound calls into VoIP packets.

Inbound calls are paid for by the calling party. Outbound calls over the PRI are billed at whatever rate the carrier charges. This is a good example of VoIP convergence: the packet-switching of the VoIP network operating with the circuit-switching of the PSTN to support both inbound and outbound telephone calls.

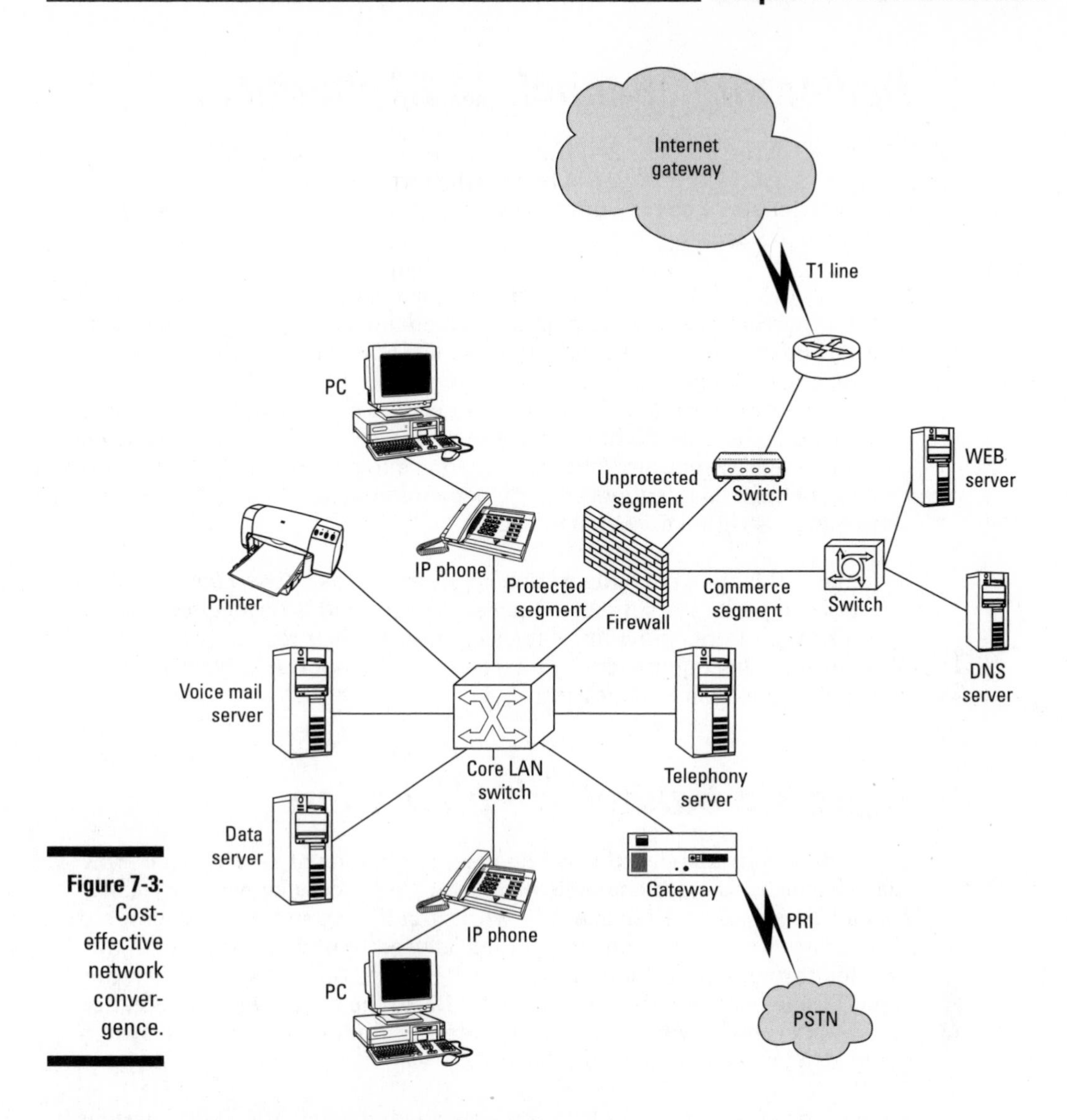

Figure 7-3: Cost-effective network convergence.

Managing Bandwidth

As mentioned, dedicated transports are channelizable, meaning that their total bandwidth can be divided by channel and assigned to support specific applications (computer data, voice or telephony, and videoconferencing). Each dedicated transport is standardized to have a preset number of digital service bandwidth channels.

Dedicating channels to applications

As defined in the DS CSI, each transport channel is called a DS0 and provides 64 Kbps of bandwidth. A *DS0* is the smallest unit of bandwidth in the dedicated carrier services network. All dedicated transports are based on some multiple of DS0s.

For any dedicated transport, the transport line can be programmed to deliver its total aggregate bandwidth capacity (all channels). Or the channels can be divided and assigned to specific applications. For example, a T1 line has a total aggregate bandwidth capacity of twenty-four DS0 channels. In a small local area network running VoIP that connects to a larger wide area network running VoIP, the LAN might optimize their dedicated channels by assigning eight channels to support VoIP, eight channels to support their data network, and eight channels to support their videoconferencing system. Figure 7-4 shows an illustration of such a network.

Sometimes this type of configuration is referred to as a *fixed-channel* solution. The T1 transport enters the company's premises and is terminated on a multiplexer. From there, individual cables connect to their respective application's terminating equipment. But fixed-channel is not the only option for channelizing dedicated bandwidth.

Dynamic bandwidth allocation

The dedicated transport can also connect to equipment such as an IP-PBX that assigns bandwidth channels on demand, a process known as *dynamic bandwidth allocation.* Here is how it works: An IP-PBX supports an onboard T1 interface module. A dedicated T1 line is used to physically connect the location through this T1 module on the IP-PBX. At the carrier's facility, the same T1 line connects this location's IP-PBX to the customer's private, dedicated WAN running VoIP. (A multiway gateway, router, or level-three switch could also be used in place of the IP-PBX.)

Whenever anyone on the LAN side makes an on-net call, the IP-PBX assigns it a single DS0 channel from the channel pool. When the call ends, the system returns the DS0 channel to the pool. The VoIP system at this location is capable of bringing up and maintaining twenty-four simultaneous VoIP calls.

Based on historical PSTN standards, one POTS line can fully support the calling pattern needs of six to eight people, on average. It is no different in the VoIP world. A VoIP channel can support six to eight people, on average. So a T1 transport used for VoIP can support 144 to 192 persons (6 x 24 to 8 x 24).

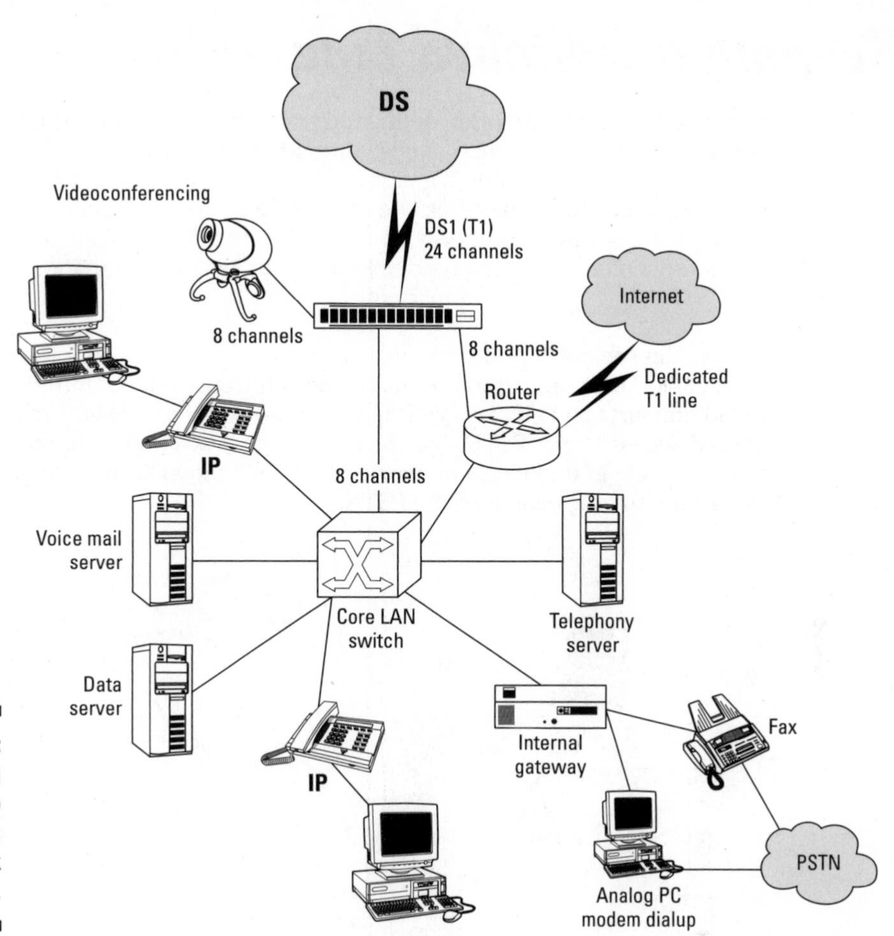

Figure 7-4: Allocating channels to specific network purposes.

VoIP enables dynamic bandwidth allocation and therefore optimizes bandwidth utilization. This is not the case in traditional networks that run computer data and telephony on separate networks. In traditional networks that use dedicated transports, it is well established that about 60 to 70 percent of dedicated network bandwidth is not even used.

Dynamic bandwidth allocation ensures that your network gets the bandwidth it needs when it needs it, and it enables a company to manage its bandwidth to minimize cost and increase productivity. It is a major reason why companies can add VoIP to their current networks using the dedicated transports already in place.

Keeping a Switched Line

In fully converged networks, dedicated transports are used for integrated operation of computer data, voice, and video applications. At the same time, there is a role for switched transports in the VoIP network. Note that when it comes down to the local level, the telephone call must ultimately go off-net to reach local telephone numbers connected to the PSTN. Figure 7-5 illustrates an example of a fully converged network with switched transport access to the PSTN.

As you can see, switched transports are used on this network to support local calls. Also supported are fax services into the PSTN. Until the entire world converts to VoIP, any VoIP network, however extensive, needs at least limited access to the PSTN. For the time being, switched transports continue to have a role in dedicated networks, and VoIP networks must make provisions for carrying calls into the PSTN.

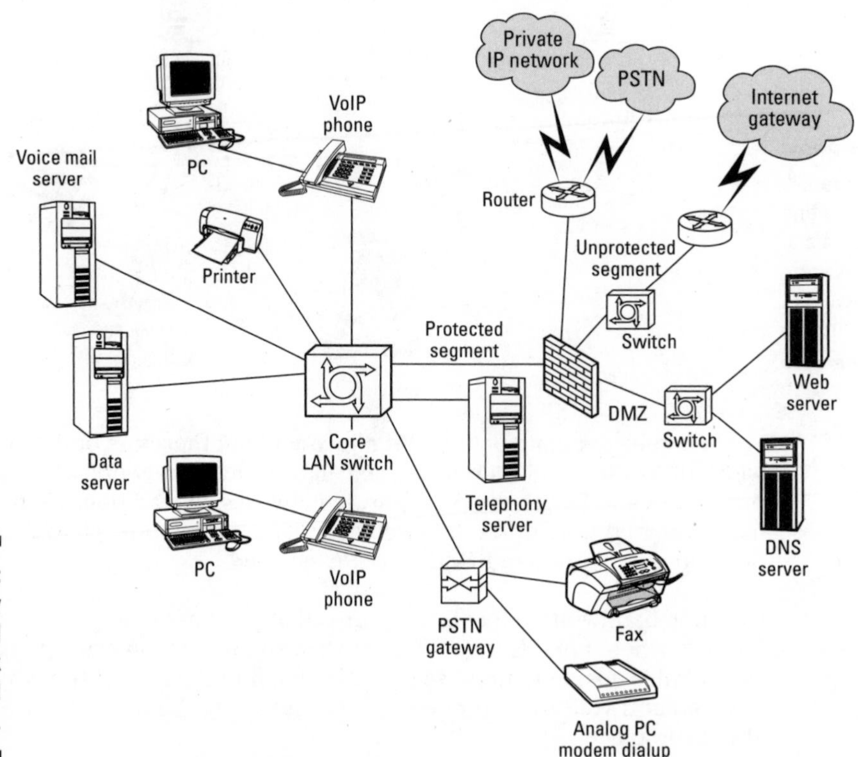

Figure 7-5: A fully converged network with PSTN access.

Another important consideration for the use of switched lines is for emergency services. The conversion of 911 emergency call centers to support VoIP has only just begun. The 911 network evolved in the PSTN and continues to be largely circuit-switched. This means it is completely dependent on the PSTN to operate effectively. Therefore, your VoIP network needs to have the ability to make circuit-switched 911 calls.

The good news is that the easiest solution is to have a PSTN gateway server (see Figure 7-2) on your network. The server simply converts on-net VoIP calls to circuit-switched calls and pushes them onto the PSTN. The cost of the gateway and switched line is recovered in the short term from your savings with VoIP.

Chapter 8

Going Wireless

In This Chapter

- Determining the benefits of wireless
- Assessing the WiFi option
- Taking wireless VoIP to higher bandwidth levels
- Getting familiar with session initiation protocol

With VoIP available on all hardwired networks (broadband and dedicated lines), you might guess that VoIP could not possibly be available in any other form. Guess again! Wireless networking not only supports VoIP but embraces it completely. This is made evident by the extent to which the wireless infrastructure provides access to each of the CSIs detailed in Chapter 4.

This chapter discusses the effect of wireless networking on VoIP, and vice-versa. Because the wireless CSI (introduced in Chapter 4) is the newest of the five CSIs and the one undergoing the most change these days, it's easy to get confused about what it provides. It includes traditional cell phone, wireless broadband, and even telephone services over your pocket PC or PDA. The good news is that VoIP works with all these technologies, but the information here focuses on how you can use VoIP with existing and future wireless networks. Chapter 10 provides more detail on using VoIP with wireless telephones.

Why WiFi?

It is fairly safe to say that wireless networking has caused a revolution in small networks and among the ranks of those perpetually on-the-go. Wireless networks allow computers to be positioned without regard for wires and make movement of computers to new locations painless. For those who are mobile, wireless networking means that they can easily connect almost anyplace they go.

At its simplest, WiFi networking is a method to extend Ethernet protocols over a limited area without the need for wires. Instead, the wireless signals become the medium by which communication happens. Before you can fully understand wireless (and how it relates to VoIP), you need to understand a bit about how wireless works. The following sections describe the basics of wireless networking.

Ethernet networking and VoIP

Ethernet comes in several flavors. When first standardized back in the mid-to-late 1980s, Ethernet was strictly a computer-data network technology that operated at 10 Mbps. Today there are 100 Mbps, 1 Gbps, and even 10 Gbps speeds. To support VoIP, each location in the network must use Ethernet as its network protocol. The network type may change as the VoIP packets traverse the distance between caller and receiver, but the network must be Ethernet at the point where the caller connects.

Ethernet, including its wireless forms, uses the Media Access Control (MAC) frame for carrying LAN traffic. The MAC frame is a way to organize the data, voice, or video bits so that they can be transported on the local network. Whenever the network traffic needs to go from the LAN to another location or to the Internet, the MAC frames on the LAN get repackaged through the LAN's edge device as VoIP packets. As packets, they can travel either on the WAN or to another network to reach their destination.

Today, thanks to VoIP, wireless forms of Ethernet such as WiFi can support voice and video applications as well as data across multilocation networks.

Examining the IEEE 802.11 standard

WiFi was invented in the 1990s as a wireless method for connecting to the LAN. It was accepted by the IEEE in 1997 as the 802.11 standard. (The IEEE, or Institute for Electrical and Electronic Engineers, is the organization responsible for the set of networking standards known as the 802 series.)

WiFi is basically wireless Ethernet. The first version of 802.11 operated at 2 Mbps and supported only computer applications. Very much like the original Ethernet 802.3 standard, the 802.11 standard evolved to incorporate ever-faster speeds. Today, variations of 802.11 include the following:

- **802.11a:** Operating at 54 Mbps, 802.11a is considered the hands-down favorite for IP telephony within a limited range. Although 802.11a is the wireless option with the best quality, it has the shortest distance limitations; you can use it up to 100 feet without the need to be connected via some wire.

- **802.11b:** Operating at 11 Mbps, 802.11b operates up to 300 feet without the need for a wired network connection. 802.11b works great for coffee shops or even small campus-type environments. Voice quality over 802.11b is passable, almost like a long-distance cell phone connection.
- **802.11g:** Operating at 54 Mbps, 802.11g is still relatively new, but is being touted as a high-speed replacement for 802.11b.

Notice that both 802.11g and 802.11a can transmit data at 54 Mbps. 802.11g has technical advantages over 802.11a, however. It is backwards compatible with 802.11b, which means that if you have an 802.11g network device, it will work in an 802.11b network (or vice versa). Thus, 802.11g provides a clear upgrade path for older 802.11b users, whereas 802.11a does not.

Newer, cheaper, faster, and better wireless transport alternatives are in the works. They follow the technology milestones completed by the 802.3 and 802.11 series of standards.

Moving up to wireless

Wireless networks take the idea of network access to a new, never-before-seen level of service, allowing more flexibility in how users may connect to the network. If your organization has already migrated to IP telephony, the LAN side of the enterprise (or the LAN side of each site location in a multilocation company) is running a standard Ethernet-based network. Because the LAN architecture is Ethernet, it is based on the IEEE 802 series of standards and is therefore compatible with WiFi.

To upgrade an Ethernet LAN running IP telephony to support wireless telephony, the only requirement is to add VoIP-compatible wireless access points (WAP). The WAP devices have a limited range, so they should be added to the network in a manner that is most useful for wireless VoIP users. All WAP devices should be connected to the LAN through switched ports as opposed to simple hubs. A hub merely provides a physical, plain-vanilla connection to the network, whereas Ethernet switches provide fault isolation at every port and therefore are more conducive with the wireless telephony application.

Finally, users need a wireless IP telephone. Such a phone looks a lot like a cell phone and operates within a few hundred feet of the WAP devices. However, users can avoid the expense of a separate phone by using VoIP soft phone on their WiFi-ready computers. Any computer that permits you to add a WiFi card can run VoIP. Wireless IP telephones and soft phones are discussed in more detail in Chapter 10.

Adding VoIP to the Wireless Network

The 802.11 standards make VoIP and IP telephony possible with nothing more than your laptop armed with a wireless interface card, perhaps a set of headphones, and IP telephony software from your VoIP provider. Theoretically, you can walk into any coffee shop or hotel that provides wireless Internet access and make a telephone call while you're surfing the Web. There are many more scenarios for using wireless VoIP, but I think you get the picture.

The following sections provide an overview of some of the VoIP telephony devices you can use with a wireless network.

IP soft phones for pocket PCs

You already know that VoIP uses packets to transport phone calls from sender to receiver. Thus, any computer that uses IP protocols can theoretically connect to a VoIP network and make or receive telephone calls. This includes pocket PCs.

Many pocket PCs running the Windows operating system can connect to local area networks and exchange data. Provided the data rate for your pocket PC is high enough, you can use the device for your phone calls. The software used on the wireless pocket computer to make and receive phone calls is referred to as an *IP soft phone for pocket PC.* In most cases, the pocket PC can also operate as a cellular phone. This makes the device versatile in a mobile, wireless environment.

Wireless extension to cellular

When using a land line, the caller pays the regulated recurring per-minute charges involved with the call. The caller has to pay these charges whether the call is answered or goes into voice mail. *Wireless extension to cellular* (WEC) is a technique that enables an incoming call to ring at your hardwired telephone (land line) and your cell phone at the same time. It enables you to take the incoming call or send it to voice mail wherever you are at the time of the call. Plus, the call does not get billed to your cell phone plan. It is treated as an inbound call to your office number.

WEC requires the PRI transport service, as discussed in Chapter 5. The PRI transport line has proven to be a cost-effective way to support WEC because it is directly compatible with the PSTN. The PRI line terminates on a PRI interface card, which is installed in the customer's premise equipment, such as a PBX. (Chapter 11 provides more detail on PBX telephone systems.) When implementing WEC, all desk phones and cell phones must be digital, not analog.

The network administrator sets up each internal extension that has the WEC feature operating on it. Each WEC user has the ability to turn WEC on and off. WEC is not a feature that a company should give to every extension because the PRI has a limited number of channels. Typically WEC is given to critical personnel who need to be reached anywhere, anytime.

After the administrator sets up an extension to have WEC, he or she provides the user with an access code for enabling or disabling the WEC application via the desktop phone. Calls coming into the company's telephone system ring at the internal desktop extension and simultaneously go out through the PRI to the PSTN. From there, they are switched to the wireless cellular network. The cell phone rings. All standard cellular phone features such as caller ID, call waiting, and voice mail are available. The cell phone is treated as a local extension of the in-house telephony system, even though the call is routed off-net through the PSTN and wireless CSIs.

Taking VoIP to the WiMax

WiMax is essentially a wireless technology that allows Ethernet connectivity over long distances — up to 30 miles. This characteristic means WiMax is likely to be used in the implementation of WANs or as the local loop for a dedicated line (see Chapter 7).

VoIP has no problem running over a WiMax network, and no special hardware or software is needed. From a VoIP perspective, WiMax is another form (wireless though it may be) of access to the wired, packet-switched network.

Graduating to IEEE 802.16

WiMax is short for Worldwide Interoperability for Microwave Access. Like WiFi, WiMax specifications are covered by an IEEE standard: 802.16. WiMax is demonstrating speeds in excess of 70 Mbps, which is more than six times the maximum speed of WiFi's 802.11b implementation.

As mentioned, WiMax covers a distance of up to 30 miles, which is a large enough range to span a city the size of Chicago. The maximum range with WiFi is several thousand square feet, depending on which of the 802.11 standards you're using. WiMax is being installed in a limited number of regions as a replacement for smaller WiFi *hot spots* because they cover a larger geographical area.

Putting WiMax to use

The evolution of WiMax has ushered in a new way of doing wireless. Although WiMax has not reached the marketplace to any great degree (as of this writing), manufacturers and service providers are developing models for how to design and sell or lease wireless networks using WiMax technology. The increased bandwidth and flexibility offered by WiMax are influencing how the marketplace views wireless telephony.

When the 802.16 standard was formalized in 2001, it quickly became apparent that WiMax had the potential for seriously changing the way we establish networks in metropolitan areas. The traditional model for network development is to install a local loop at each site location (LAN) to be connected to the larger overall network (WAN). Instead of running a T1 local loop to each location (at an expense averaging $450 to $1200 per month per location), you could have one WiMax hub that wirelessly interconnects all four locations.

If you run VoIP over this WiMax network, you also reduce the cost of intersite telephony to $0. You can also reduce the cost of your off-net regional (intralata) telephony to the price of local calling. All VoIP on-net traffic would be carried on-net to the closest of the four LANs, and there it would go out to the PSTN as a local call.

If your company has other locations that are beyond the range of your WiMax infrastructure, you can interconnect them by connecting all the WiMax hubs at each metropolitan area. You can then run traditional dedicated transports between the regional WiMax hubs.

For example, suppose you have a WiMax hub covering four offices in Chicago. You could connect this hub via a dedicated line to a similar WiMax hub that covers your offices in Los Angeles, Pittsburgh, or any other metropolitan area. Figure 8-1 shows how such a network would look if you didn't use WiMax. Figure 8-2 shows the same network redesigned to take advantage of WiMax.

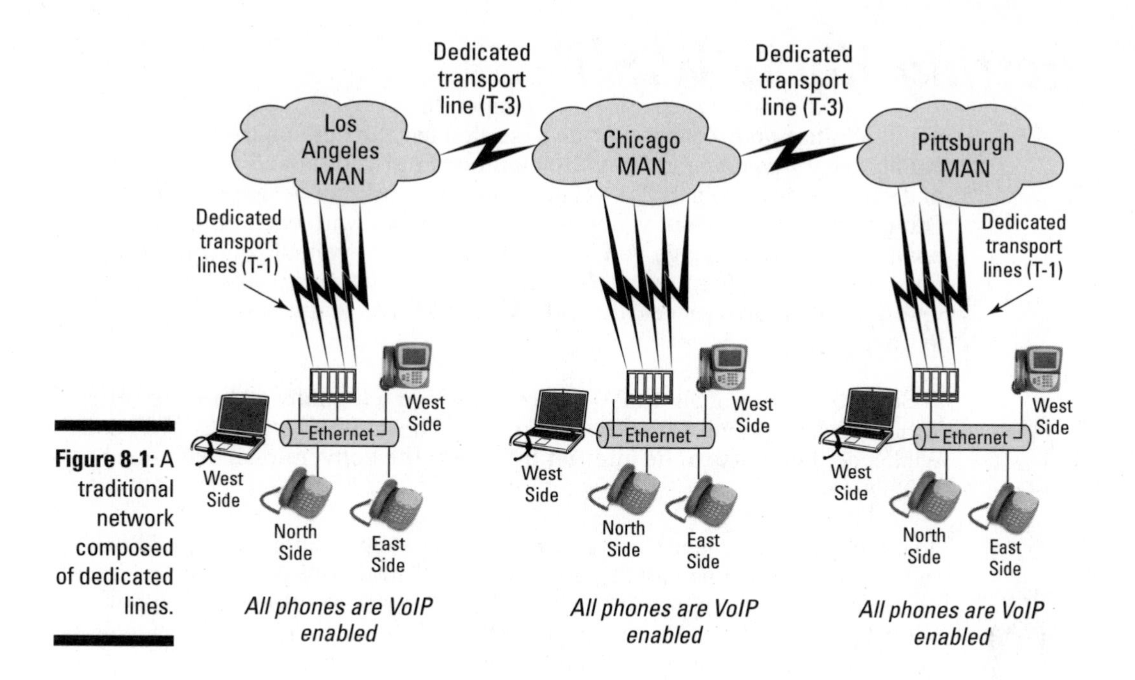

Figure 8-1: A traditional network composed of dedicated lines.

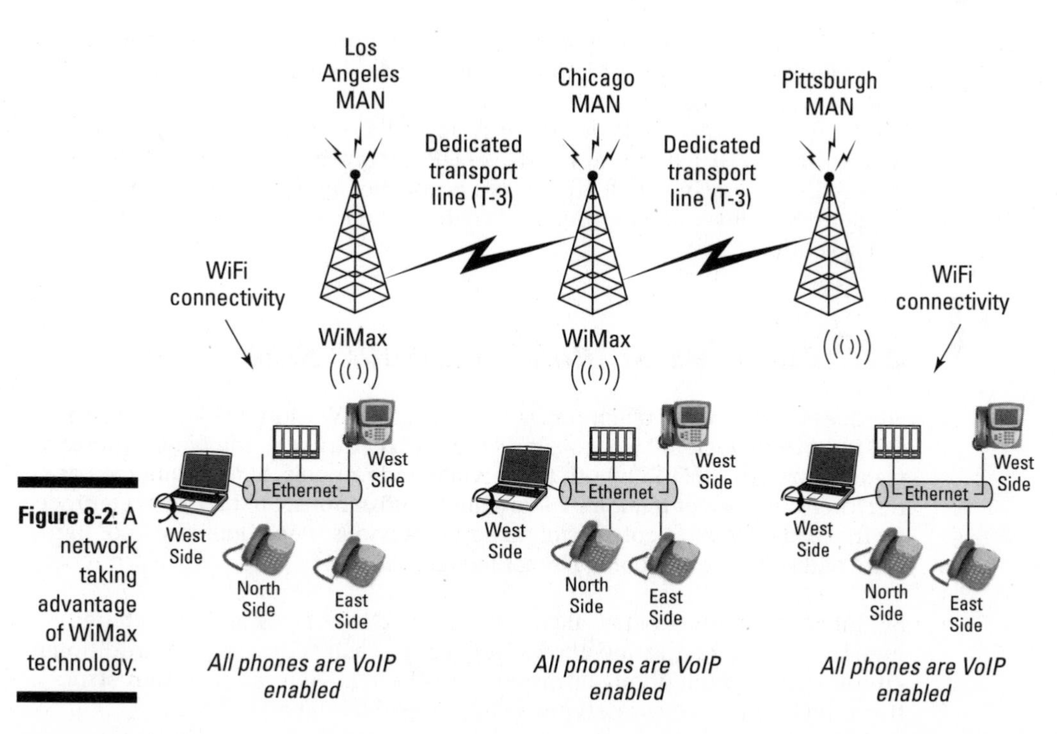

Figure 8-2: A network taking advantage of WiMax technology.

Getting Hip to WiSIP

Session initiation protocol (SIP) is a catalyst for the next phase of open communications using not only IP telephony and VoIP but also the full suite of IP-related protocols. SIP is an interoperable protocol designed to allow equipment from different vendors to communicate with each other. SIP enables new leading-edge VoIP calling features not available on traditional phones, including features such as LDAP directory access, presence, and multiple call appearance. (You find out about these features — and more — in Chapter 10.)

WiSIP is the flavor of SIP that delivers advanced telephony features to other WiSIP wireless phones or end-user devices. These devices can use WiFi and WiMax connectivity to the Internet to bypass the conventional cell phone and PSTN networks.

It is important to realize that you do not need a WiSIP phone to benefit from the advanced features of VoIP through the SIP protocols. A VoIP end-user with multiple endpoint devices such as a cell phone, a desk phone, a PC client, and a PDA can rely on SIP to permit all these devices to operate as a single system. SIP brings about increased efficiency and productivity.

In a VoIP converged network with SIP, organizations can choose from a variety of vendors to create a seamless converged communication network. For instance, some equipment vendors use SIP to support trunking functionality. (*Trunking* is a way to make a network support a protocol it might not otherwise support.) Other vendors may use SIP to control gateways and calling features. The way in which SIP is used is entirely up to the communications vendors. SIP allows their equipment to communicate with equipment from other vendors.

SIP enables smoother conversions

SIP does to traditional telephone service what the World Wide Web does to the Internet. In fact, SIP is a cousin of the main protocol of the Web, hypertext transfer protocol (HTTP); both are text-based protocols. SIP has emerged at the forefront of most if not all VoIP-related applications. SIP has been embraced by the leading VoIP telephony manufacturers and is being built into VoIP hardware and software, including IP-enabled telephones.

SIP integrates with traditional circuit-switched interfaces and IP-switched interfaces. This integration allows the user to easily convert from traditional circuit-switched telephony infrastructures to next-generation IP infrastructures, including wireless networks supported by WiFi and WiMax.

Using SIP today

Wireless SIP telephones enable the user to place telephone calls through any WiFi hot spot, over the Internet, to anywhere in the world. The call bypasses the PSTN; there are no fees, recurring charges, or any costs associated with the call, except perhaps a charge for gaining access to the WiFi network. (You could always go to another WiFi hot spot where there is no charge for access.)

If you use a WiSIP phone, there are absolutely no carrier charges no matter where in the world you call. The downsides are the cost of the phone itself and the fact that whoever you call with a WiSIP phone must also have a WiSIP phone. In the future, this will undoubtedly change to enable any WiSIP phone to call any wireless device regardless of type, provided the receiver supports VoIP protocols.

Chapter 9

Using VoIP on the Internet

In This Chapter

- Understanding how your choice of network affects QoS
- Examining protocols
- Securing your network with firewalls
- Making the Intranet accessible
- Connecting with a virtual private network

The Internet is often referred to as a network of networks. The fact of the matter is that the Internet is just another network, albeit with an enormous and global reach. You can access the Internet through the public switched network (PSTN), your wireless personal digital assistant (PDA), or your WiFi notebook at the coffee shop nearest you. You can use your broadband link to access data, voice, and video at home. Your company can access the Internet through any of the dedicated transport lines. (Chapter 4 covers all Internet access network types.) Slow-speed, high-speed, or ultra-high-speed avenues of access can all work simultaneously to support millions of users accessing some portion of the Internet somewhere in the world.

No other network in the world has such a level of accessibility. But this accessibility also makes the Internet inherently less secure than other networks. Whenever a company establishes a connection between their dedicated network and the public Internet, they open their dedicated network up to many of the same security hazards endemic to the Internet.

Regardless of the hazards, many companies need to provide Internet access to their employees and open at least a part of their private network to the public through the Internet. Companies establishing Internet connections need to be concerned about security — period. Corporate sabotage, identify theft, and denial of service attacks (common on the Internet) have generated billions of dollars in losses. Although these threats pertain largely to computer data applications, and not VoIP voice packets, anything that negatively affects the overall network can also affect VoIP applications. There is no doubt that corporate use of the Internet has raised the bar on network security.

This chapter discusses two primary issues related to VoIP and the Internet. First, it discusses using VoIP over the Internet, particularly the quality of the VoIP connection over the Internet. The second issue is related to security: connecting your VoIP-enabled network to the Internet and still ensuring the safety of your servers and data.

Network Options Affect Quality of Service

Running VoIP over the Internet is an option, but it is clearly not the best option for most companies. The goal of any VoIP implementation is quality of service, and the key to achieving that quality is figuring out what is best for your specific network configuration.

Most companies install an intranet to facilitate internal access to applications and data. An *intranet* is a network that uses the same protocols and tools as those used on the larger Internet, such as e-mail, Web browsers, and instant messaging. The difference between an intranet and the Internet is that an intranet is private, typically created for the access of employees and selected vendors or customers, whereas the Internet is public and open to anyone.

You do not need a connection between an intranet and the Internet, but if your company chooses to create a connection, it uses a firewall with a gateway at the connection point. These allow a company to monitor and secure traffic over the connection. With a firewall and gateway in place, a company protects itself against Web-related traffic affecting their private, dedicated network — including their VoIP network.

Some small, single-location companies that cannot afford their own private wide area network use the Internet as a medium for transferring VoIP traffic from their LAN to the outside world. If voice communications are critical to this business, I don't recommend running VoIP this way because the quality of the voice calls is at the mercy of factors out of the company's control. It is better (and more secure) to use a VoIP hosting company. This separates VoIP traffic from all other Internet traffic while keeping the rest of the network in place.

Large companies with many sites often use private, dedicated transport lines to ensure the requisite quality for their VoIP traffic. The network uses high-bandwidth connections between major company facilities, with smaller-bandwidth connections to the smaller company sites. In companies with a myriad of site locations, it is common to establish large regional hubs for their networks. Each smaller site connects to its respective regional hub using dedicated transport lines, sized according to their bandwidth needs.

Provided that the company's dedicated network has enough bandwidth, intranet traffic can also be run over the same connections. If Internet access is needed over the company's network, they often establish an Internet connection at one of their facilities (such as company headquarters) and then use that connection to provide Internet access to remote sites on the company network. This type of access plan provides a way for companies to limit the number of firewalls and gateways necessary, as well as minimize Internet access fees.

Figure 9-1 illustrates a typical regional hub for a company running both VoIP and an intranet. VoIP calls get carried on the private dedicated network. The private network is physically separated from the corporate Internet connections and, therefore, the actual Internet. Notice that the firewall controls all traffic from the Internet for security purposes. It also controls access for nonemployees accessing the corporate intranet through the public Internet. (Firewalls are discussed later in this chapter.)

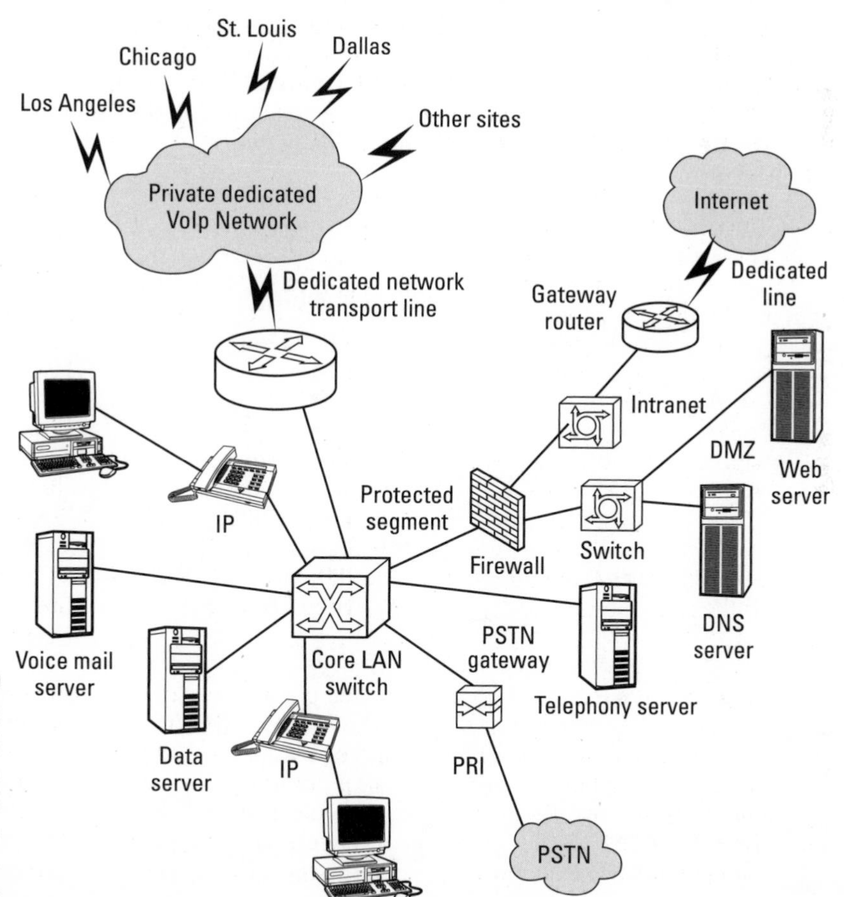

Figure 9-1: Regional hub for a large multisite network.

Internet Protocols and Quality of Service

As discussed in Chapter 1, TCP/IP uses a layered approach to networking. When VoIP is the network transport service, the UDP protocol is substituted for the TCP protocol. UDP (user datagram protocol) is one of the many protocols included in TCP/IP, so VoIP can be made to run on any network type.

The Internet is not the only network that supports VoIP. Any network — private or public — that runs the TCP/IP protocol suite can run VoIP. Quality of service varies from network type to network type, and the ISP you choose can affect the QoS you experience.

ISPs make the Internet go round

An Internet service provider (ISP) is a company in the business of providing Internet access to consumers and businesses. It is common to rank the quality of ISPs based on their tier level (see Figure 9-2). Tier-0 is a logical ring formed by all the Tier-1 ISPs. Tier-1 ISPs are considered the largest and usually the best type of Internet access because there is only one "hop" between the tier-1 ISP network and the end-user's network.

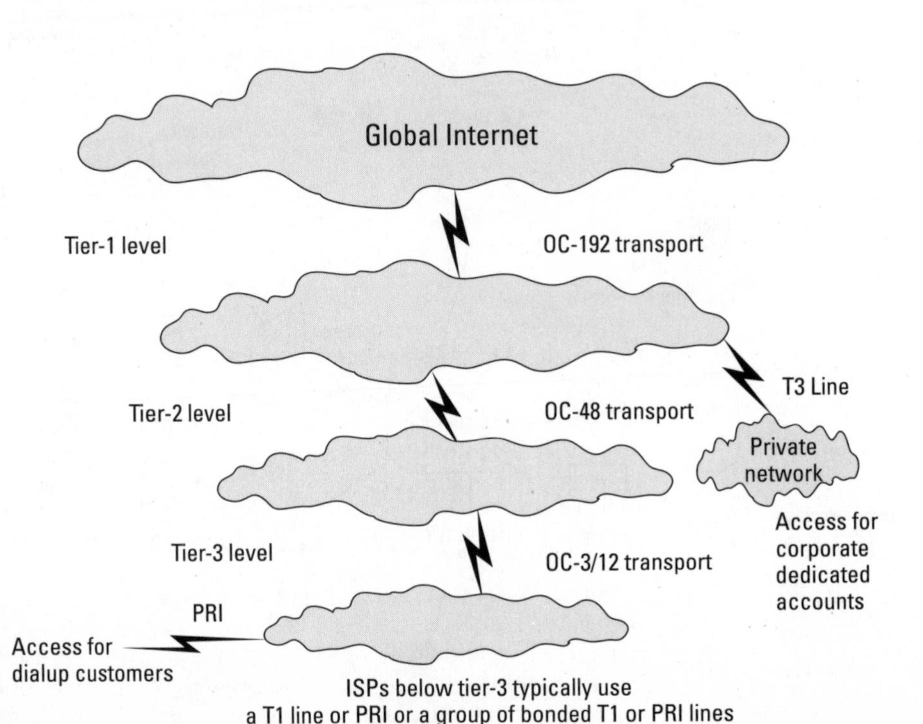

Figure 9-2: The tiered nature of the Internet.

In geek speak, a *hop* is a connection between networks. Thus, if a packet travels from your network to your ISP's network, that is one hop. If the packet then travels to a larger carrier (perhaps someone from whom your ISP leases lines), that is another hop. It is not unusual for packets traveling through the Internet to go through fifteen, twenty, or more hops from source to destination.

National and international ISPs are all tier-1 and typically have multiple, ultra-high bandwidth pipes connecting them into the core of the Internet. Regional ISPs tend to be tier-2. Smaller ISPs that acquire their Internet access through a tier-2 ISP are tier-3 ISPs. A few ISPs operate at the tier-4 and tier-5 levels.

When it comes to VoIP, each hop adds transmission overhead that may diminish the overall throughput of the call. On the one hand, if the network types involved in providing the end-to-end connection are using strictly dedicated bandwidth transports (see Chapter 7), you may not notice degradation. On the other hand, if you are using an ISP (regardless of their tier) that uses switched transports somewhere in the end-to-end connection, you are going to experience some degradation. This is because switched lines do not pass packets as well as dedicated lines. (That is why VoIP networks are designed using dedicated transports for on-net traffic and use switched lines only when it is necessary to go off-net to the PSTN.)

Depending on the number of hops over the Internet and the types of network lines between the VoIP caller's network and the VoIP receiver's network, there may be delay; it is inevitable when using the Internet. Having multiple hops does not necessarily equate with a poor connection, but it can be a significant factor.

How you choose to access the Internet and whether you then use the Internet to support VoIP are choices you need to make. In general, for companies with more than four or five locations, the Internet is not a good choice for VoIP, although it may be a good choice for transferring computer data.

Examining protocol layers

You already know from Chapter 1 that VoIP runs with and requires the use of TCP/IP protocols. The good news is that the way these TCP/IP protocols work for data networking is the same way they work for VoIP, with the exception that VoIP utilizes UDP instead of TCP and also requires that some additional information be packaged with the data packet.

In the process of being transmitted from source to receiver, VoIP data follows the same process to construct and transmit packets as is followed by other TCP/IP data packets, as shown in Figure 9-3. The difference is that UDP is used instead of TCP at the transport layer; this is probably the most significant difference between computer data packets and VoIP telephony packets.

"Hello! how are you?"  NTP - RTP - RTCP	Application	Surfing the Web  PC HTTP
UDP packet part	Transport	TCP packet part
IP address packet part	Internetwork	IP address packet part
MAC address packet (frame) part	Network interface	MAC address packet (frame) part
Physical electronic signaling outbound	Physical	Physical electronic signaling inbound

Figure 9-3: Differences in TCP/IP implementation for VoIP and data packets.

Because VoIP packets are constructed pretty much the same way as data packets, VoIP can run on any type of data network that utilizes TCP/IP. Because most corporate networks already handle TCP/IP traffic, the fact that VoIP packets can travel on the same network means that converting to VoIP can be relatively painless.

Firewalls for Security

Many of the security measures devised since the early 1990s have been centered on protecting private networks from the public Internet. Today the larger corporate networks use a firewall when connecting their private networks to the Internet. They then can run VoIP on their private network without threat from the Internet.

In the same way, smaller companies with fewer than four or five locations can elect to go with a private network in a similar way. But depending on their business needs and total number of employees, it may make more sense to use a virtual private network (VPN). VPNs present interesting firewall security challenges. (More on VPNs in a virtual minute.)

Consumer VoIP and firewalls

Consumers running VoIP over their broadband service have little to worry about with Internet security. No doubt someone will try to sell you a firewall security package to protect your VoIP service from a virus attack, but it would be foolish to spend the money.

Broadband providers and VoIP carriers include with their services a first level of firewall security. Your VoIP calls have firewall security first from your VoIP carrier and then again when your voice packets travel through your broadband provider's network.

Does this VoIP protection keep you safe if you download unsafe data files to your computer? No! You still should have a firewall security package running on your computer that protects you against nasty, corrupt, data Internet files. You don't need such protection for your VoIP calls, however.

Most of the security measures utilized today read and inspect the packets that traverse the network. Security therefore comes down to deciding where in the network the packets will be examined and what type of systems you will use to monitor the packets. For most companies, the examination location is at the connection point to the Internet, as described earlier in this chapter.

Companies that implement a VPN require each location on that VPN to use a firewall to access the Internet. If the company is to run VoIP over the VPN with good QoS, each site must use private, dedicated transport lines to connect their respective locations to the Internet.

The firewall must be configured to support tunneling of VPN traffic. *Tunneling* is a method for ensuring that packets traverse the public Internet in a secure manner that prevents disruption. With this type of network infrastructure, each location's LAN is secure from outside Internet threats and VoIP telephony traffic is carried over the Internet.

Three major categories of firewalls are available. Each comes with a different price tag and level of sophistication or complexity. Depending on the firewall, a company may need to have full-time skilled staff or corresponding contractors available to set up, program, and maintain the system.

The three types of firewalls follow:

- **Packet filter:** This is a simple firewall that entails only nominal cost. Some are even free. Consumers, mobile users, and very small businesses use the low-end flavors that cost from $0 to $59. Some small companies use router devices that come preprogrammed with this level of firewall security and cost from $800 to $2500.

- **Proxy server:** Full-fledged proxy servers tend to be complex to establish and maintain. They are more suitable for use at a single location. Because they must first permit unwanted packets on the network before they can discard such traffic, they are controversial in multisite networks. Costs range from $5000 to $15,000.
- **Stateful event monitor (SEM):** These firewalls are the most comprehensive and therefore the most complex. They are typically used in large networks that have an inside protected network that needs to be separated, at the physical layers of the network, from their public Internet access network and their semipublic intranet network. SEM firewalls can examine the entire contents of any packet it sees, make decisions based on any field contained in the packet, and, like any firewall, can reject any packet from access into the company's protected network or even to its intranet. SEMs include extensive network management capabilities. Costs range from $24,000 to $35,000.

Figure 9-4 illustrates how a SEM firewall does its duties. This is the same type of firewall used in large networks, such as those shown in Figure 9-1. Most large corporate networks don't use VoIP over the Internet. VoIP packets instead traverse the company's private dedicated network and never even see the other side of the firewall.

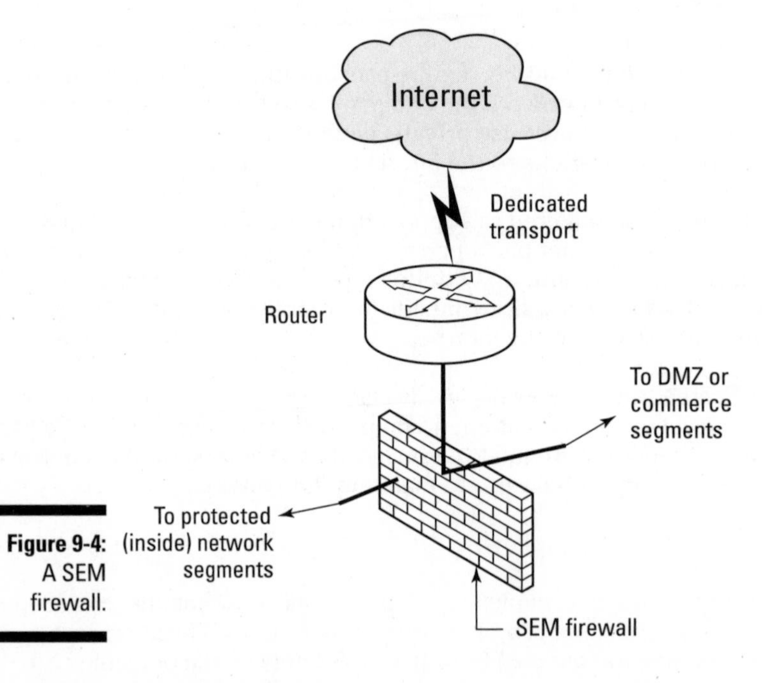

Figure 9-4: A SEM firewall.

A SEM firewall protects the network it runs on in many ways, but its main duty is to read packets that arrive at its "doorstep" and decide whether the packet is authorized to be there. If the packet is bad, the firewall discards it. If the packet is from a friendly network or user, packets that most likely are from customers and visitors, it may permit access to a portion of the company's intranet. If the packet is from an employee or other authorized person, the firewall permits access to the inside protected network, to the resources the person is authorized to have.

If you have a VoIP phone that includes Web features — such as displaying Web pages — these features require fetching data that travels through the corporate firewall. This data, however, is standard Web data; it is not VoIP packets, which remain on the dedicated network.

Connecting Through a VPN

If the Internet must be used for your company's VoIP traffic, you have only one option to consider: a virtual private network (VPN). It has proven to be both cost-effective and capable of delivering good quality VoIP service for companies with limited VoIP needs.

The term *virtual private network* is an apt description. The *virtual* component was intended to convey *virtually* anywhere. The *private* is derived from the fact that a VPN uses private dedicated transports at each location on its network to connect to each of their respective local ISPs.

The VPN concept emerged in the early 1990s as a way to transfer data securely over the Internet. Consider the case of a small company with two locations, one in New York and the other in Los Angeles. Instead of using a T1 line to connect the offices at a cost of $12,000 per month, the company would pay $1200 per month to get a T1 connection at each location from a local ISP. They would then use the Internet as the backbone network to do countless computer data applications. Figure 9-5 shows an example of a VPN that connects three locations.

Historically, the use of the Internet as a network backbone is called *extranetting,* or riding the Internet for free. In the 1990s, VoIP and security issues were not even in the picture.

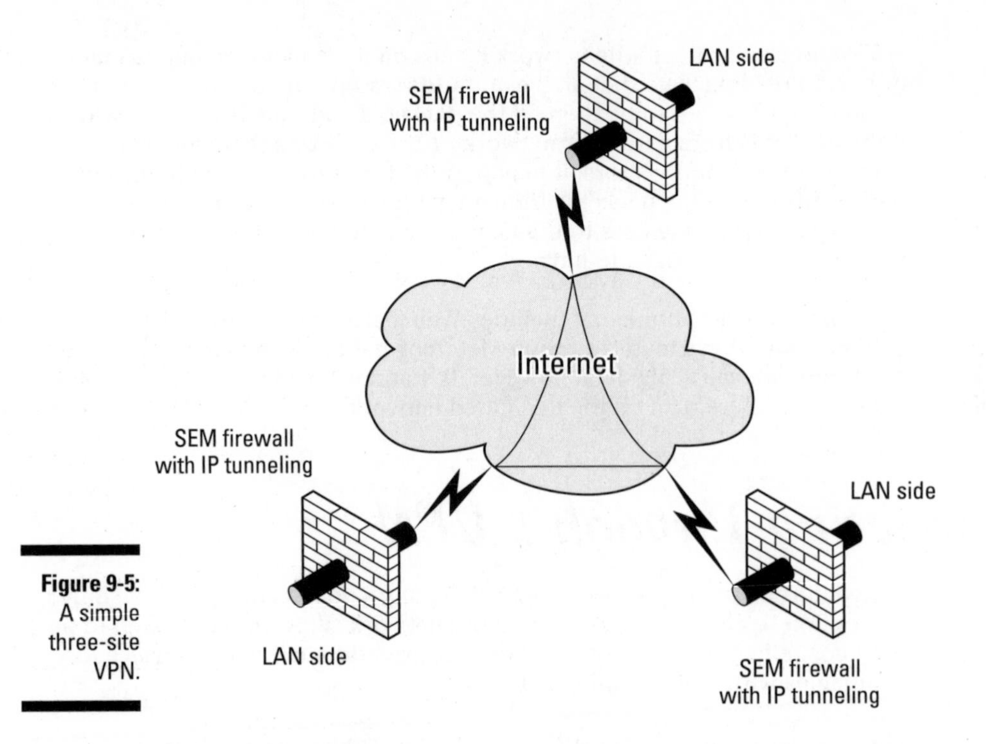

Figure 9-5: A simple three-site VPN.

Today, any discussion of using the Internet as the backbone of a private network inevitably leads to a discussion of VPN. VPNs require some sort of secure gateway, firewall, or router at each location connected to the network. Private, dedicated transport lines are used to connect each location to their local ISP. Each gateway is configured to route all traffic — including VoIP traffic — over the Internet to each of the other locations in the company, as well as to the Internet generally.

Because of the contentious nature of the Internet and the high cost of securing each VPN location's network, VPN designs are proving to work well when the network has no more than five to eight locations and less than twenty people per location. However, no significant studies have determined what the maximum permissible number of users per location should be.

Before implementing a VPN, a company must undertake a thorough analysis to assess calling patterns, call volumes, hardware needs, bandwidth requirements, Internet access, and security needs — for each location on the VPN. After your present network requirements are determined, you also need to plan for future growth. All of this would then need to be balanced against the total cost of operation and the complexity of having a VPN.

VPN costs

In a VPN design, each site bears the full cost of the following:

- Dedicated private connection to their ISP
- ISP Internet bandwidth access
- Terminating hardware including firewall and router (minimally)

These costs are much higher than a typical single site's cost to connect to a larger private dedicated network due to each VPN site's need to have their own Internet access. VPN sites also require more complex network hardware configuration. However, after the connection is set up, no other recurring costs are involved because the Internet is basically a free ride. Consequently, running VoIP over a VPN can be very cost effective if the configuration can be completed in a manner that ensures security and high QoS.

The more sites added to a VPN supporting VoIP telephony, the less cost-effective it may become. Remember that each site added requires their own high-speed access to the Internet, their own router, and their own firewall. You also must consider the administrational burden for network administrators (for managing that firewall and router). All this doesn't come cheap. Ultimately a point is reached where establishing a dedicated network becomes more cost-effective than continually upgrading your VPN. Exactly where that point is depends on many factors, but the primary ones are distance between offices and the number of offices you need to connect. Your company should do a thorough analysis to determine exactly which approach is best for your goals.

If your company has quite a few mobile or remote users, establishing a VPN may make strategic sense. Their personal computers can function as routers and firewalls, and the fact that they can connect to the company network over the VPN from any Internet access point can be a big plus. Make sure you consider the needs of your mobile and remote users in any analysis you undertake.

Implementing a VPN

VPN hardware and software technology has evolved into two distinct categories: gateways and firewalls. *Gateways* allow individual LANs to connect to the Internet. They can perform VPN-related tasks, as well, such as encrypting and decrypting data transferred through the gateway. The gateway physically connects each LAN to the transport lines used for Internet access.

Gateways include software that enables the network administrator at each site to manage the network. Anyone attached to the LAN would then be able to access the Internet-based external network and any of the other corporate LAN sites attached to the VPN. Through the use of IP-enabled telephones or digital telephones optimized to support VoIP, any user can make VoIP telephone calls to anyone at any of the sites on the VPN.

Mobile users connect to the VPN through client software (provided with the gateway) installed on their laptop. When the user is in the office, the laptop connects to the VPN just like everyone else's computer. When away from the office, the mobile user uses the client software to access the corporate VPN through any Internet access port. To support VoIP telephony, the user needs to run IP soft phone software on the computer. After obtaining access to the VPN, the user can make and receive VoIP telephone calls through the IP soft phone.

Firewalls, the other VPN category, are used to implement security on any network to which they are attached. (Firewalls were discussed earlier in this chapter.) Privacy and protection are important when using the Internet for any service, including VPN and the VoIP services that may operate over the VPN.

Part of the complexity involved with the design of a VPN is the configuration of the firewall and other computers exposed to the Internet. But this complexity enables each location on the VPN to nail down tight security. Through the gateway or through a separate additional firewall device, each site can set up protection from unauthorized intrusion.

The Internet Engineering Task Force (`www.ietf.org`) has developed the IP Security protocol suite, or IPSec. This is a set of IP extensions in the form of software, just like the entire TCP/IP suite of protocols. IPSec is installed on the gateway (or separate firewall if used) to monitor each packet that passes through. Unauthorized or questionable packets are discarded prior to entry into the protected segments of the network.

Many variations of VPNs and hundreds of VPN service providers are available. To make the best decision regarding a VPN, do your homework and investigate the options available. It is worth your time to meet with various companies and solicit bids for implementing your VPN. The more you know, the better decisions you can make.

Chapter 10

Telephones and VoIP

In This Chapter

- Summarizing VoIP phones
- Checking out hard phones for the desktop
- Using soft phones for the computer
- Choosing wireless IP phones
- Protecting your existing phone system investment

Depending on the number of people in your company, the mere thought of purchasing replacement phones could make you nervous. That's why one of the first questions people ask when considering VoIP is whether they need to buy new phones. The answer to that question rests primarily with the type of telephone system your company already has.

This chapter examines the ins and outs of telephones and VoIP. You discover the options for new equipment, as well as how you can use your existing equipment with your new VoIP system. To keep you from tripping over your tongue, I'll refer to *VoIP phones* instead of *VoIP-enabled phones.* A VoIP phone simply means a phone capable of placing and receiving calls on a VoIP network.

Running Down the Three Flavors of VoIP Phones

Unlike older telephones that must be hardwired to a PBX, VoIP phones connect directly to the LAN, just like a computer. LANs use a different type of cabling than traditional telephony systems, so VoIP phones have a built-in network interface card (NIC) that provides the connection port for the LAN. VoIP phones also have their own MAC address, which is required for peacefully connecting to an Ethernet network.

Some models of VoIP phones come with extra Ethernet ports. You plug the phone into the network, and then plug the computer into the phone. The benefit is that you need only a single network connection to connect both phone and computer. This can save your company money.

Many makes and models of VoIP telephones are available. As was the case with cell phones a decade ago, discounts, rebates, and other inducements are often offered to customers to entice them to buy VoIP phones. Some VoIP providers reduce the cost of the phone or give it away in return for a long-term service commitment of two years or more.

VoIP phones can be plain-Jane and basic, or they can be full-featured and support videoconferencing and Web surfing. You can even put free VoIP software on your computer and eliminate the VoIP phone altogether. VoIP has introduced a new way of viewing the telephone.

Following are the three distinct categories of VoIP phones:

- Hard phones
- Soft phones
- Wireless phones

Phones in these categories vary depending on many factors besides price, as you discover next.

VoIP Hard Phones

If you can see it, feel it, and tether it with a network cable, and if it includes a traditional phone keypad, you have a *VoIP hard phone.* More phones are in this category than in the other two categories combined. Because there are more makes, models, and manufacturers, competition helps lower the price.

Even though there is much diversity in this category, two things should be common to every VoIP hard phone: support of TCP/IP protocols (mandatory for VoIP) and at least one RJ-45 connection port.

The *RJ-45 connector* on a hard phone is an Ethernet port used to connect the phone to your network. Through this port, your phone can communicate with any other IP-based device on the network. These devices include servers that keep track of everybody's telephone number and voice mail, other VoIP phones, the gateway to the PSTN (for off-net calling), and the router that takes care of establishing a connection to other VoIP phones on the network (on-net calling).

The RJ-45 port looks like a regular phone jack (RJ-11), but it's a little wider. In Figure 10-1, the jack on the left is an RJ-11 and the one on the right is an RJ-45. It is the jack on the right that you would use to connect this VoIP hard phone to the network.

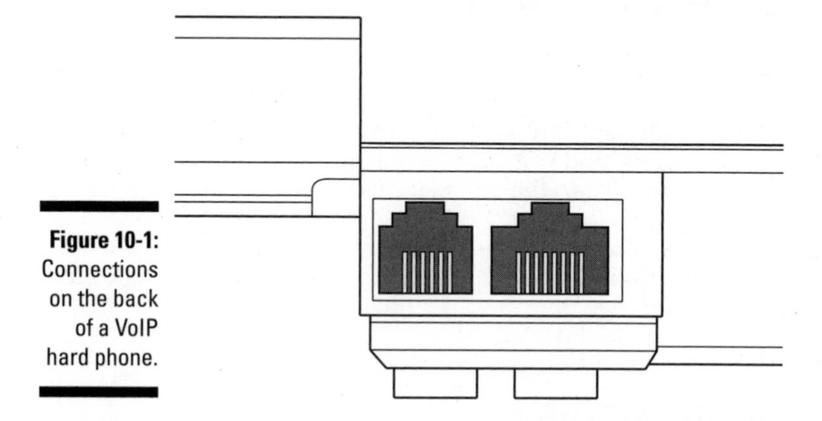

Figure 10-1: Connections on the back of a VoIP hard phone.

Some VoIP hard phones come with multiple RJ-45 ports. These phones have a built-in switch, which means the phone can be used like an Ethernet switch to connect more Ethernet devices (such as a computer or a networked printer) or even another VoIP phone.

The VoIP hard phone looks the most like a traditional desk phone. Hard phones can be broadly categorized as basic, intermediate, and advanced phones, based on their capabilities.

Basic hard phones

Basic VoIP hard phones look like a traditional desk phone. The dialing pad is clearly distinguished. This type of phone is considered a basic, entry-level IP phone that delivers good VoIP telephony service. That is, it makes and receives telephone calls over the VoIP network (on-net) or the PSTN (off-net). You could find such a hard phone on the desk of a staff person or in common areas such as the lobby or hallway of any typical company.

Intermediate hard phones

An intermediate VoIP hard phone has a large screen and many more hard buttons compared to the basic hard phone. Intermediate phones can do anything that the basic hard phone can do plus more. This phone can often do Web browsing and access the company phone directory.

Advanced hard phones

Advanced hard phones take VoIP telephony to a new level. They usually include color video displays and multiple telephony-related applications. These phones have fewer hard buttons than intermediate hard phones because the phone's screen and software can enable many more functions than could be manufactured into the phone's chassis.

Figure 10-2 shows an example of an advanced hard phone. This hard phone delivers not only exceptional services and features but also a true and complete interface to the Web.

Figure 10-2: Advanced VoIP hard phone.

Features supported

Telephony features can be delivered in two ways:

- As a function of your VoIP hard phone
- Through the VoIP network to the phone from another device attached to the network, such as a server or a telephone controller

In the older world of telephony (pre-VoIP), features were known as *call features, line features,* or *system features.* You paid for these features each month, sometimes on a per-line basis. This may not seem like much in the grand scheme of things, but if you or your company have multiple lines, feature costs can significantly increase your monthly phone bill.

In the VoIP telephony world, all features are free.

The VoIP hard phone itself plays a role in exactly what type of features you receive. Its common features (like those on traditional phones) include the following:

- Hold
- Conference call
- Transfer
- Drop call
- Redial
- Volume up and down
- Mute
- Speaker
- Messages (voice mail, including an indicator light)

The basic VoIP hard phone also provides at least two *call appearances,* the ability of the phone to bring up and maintain separate telephone calls as if you had separate physical lines. Call appearance buttons are usually located near the Hold button and labeled 1 and 2.

Intermediate phones usually include a flat screen. Some of the more expensive ones provide for limited Web browsing. These hard phones also come with the ability to receive their electric power from the network. This means the LAN can provide the power the phone needs through its network connection. (You may hear this referred to as power over Ethernet or simply PoE.) As a result, you don't need to plug in a power cord at your desk.

Traditional features found on any basic VoIP hard phone are provided by buttons on the phone. The intermediate phone usually includes several buttons, but it also has many other features provided through its software and screen. Common features on intermediate phones include the following:

- Graphical display screen
- Presence indicator lamp
- Multiple programmable feature keys
- Application buttons that parallel the screen (you can set these up to, for example, bring up your speed-dial list or browse the Web)

- Integrated switch ports for connection of the PC
- Full-duplex speakerphone (people can talk and hear simultaneously)
- Integrated headset jack
- Multiple language support
- Support for simple network management protocol (SNMP)
- Hearing-aid compatibility
- Multiple personalized ring patterns
- Voice media encryption

Some people refer to advanced hard phones as *appliances* because they do more than just allow you to carry on traditional voice conversations. For example, they usually provide Web-related features and may include other applications. Advanced VoIP hard phones include all the features found on basic and intermediate phones. With its larger video screen, the advanced hard phone represents an amalgamation of both computer and VoIP phone. If it weren't for the hard telephony dialpad buttons, you might confuse the advanced phone with a computer.

In addition to the features already listed, the advanced hard phone includes these features:

- **Phone:** Allows the advanced phone to use capabilities offered through a telephony server or other telephone system connected to the VoIP network.
- **Directory:** Provides access to the corporate LDAP (lightweight directory access protocol) server on the network. With this type of access, you don't have to even dial the number. You can look up the name on the LDAP and press one button; the network takes care of the rest.
- **Web access:** Advanced hard phones have expanded access that is closer to what you might expect from the browser on your computer. Web access capability often includes support for Java applets (self-contained programs created in the Java language).

VoIP Soft Phones

If your computer is connected to a network using TCP/IP, you have the capability to run a VoIP soft phone. *Soft* is a term that comes from past references that compare printed output (hardcopy) to screen output (softcopy). In other words, if something is based on your computer screen, it is *soft*.

Like the VoIP hard phone, the computer running the VoIP soft phone needs to have LAN access. The big difference between the two is that the soft phone has a "soft" dialing pad on the screen instead of a "hard" dialing pad on a physical phone. To dial a number, you use the mouse to point and click. Some versions offer touch-screen dialing.

Besides the need to have the appropriate audio functionality in your computer (sound card and speakers or a headset), the viability of using a soft phone really comes down to three things:

- The network
- The quality of the software
- The type of computer you're using

If you are in the corporate world, the *network* is already established. There are options if you are mobile and working on a laptop, but your workplace is still managed by the company. At home or for consumer use, the Internet is your network.

The *software* is the application that interfaces with the network using the TCP/IP protocols required by VoIP. At work, your company provides the software. For home or leisure use, you can program your own soft phone software or download one of the many packages now available on the Web. The going rate for these plans start at about $30 per month.

Be careful not to give out your credit card information unless you are ready to sign up with a VoIP provider, and be particularly wary of giving information to any third-party who says they are selling VoIP services. As with any transaction of this manner, it makes sense to safeguard your financial information as best you can.

One benefit of soft phones versus hard phones is that the soft phone doesn't have to be connected to the network using a cord. Soft phones can run just as well on a computer that supports wireless (WiFi) networking as they can on a computer that is connected to a network through a cable.

That leaves the third element: your *computer*. Fortunately, you don't need a lot of computer power to run a soft phone. If the computer can connect to the network, it should support a soft phone. There are two categories of computers: stationary and portable.

Stationary computers

VoIP soft phones do not necessarily have to be on a laptop, a notebook, or a tablet PC. These phones run just as well on a stationary (desktop) computer. Depending on your needs, a soft phone can eliminate the costs associated with a hard phone.

Also, VoIP soft phones can support videoconferencing. If you need to run video, the computer screen has an obvious size advantage over a video phone's screen. Figure 10-3 shows a soft phone on a computer screen.

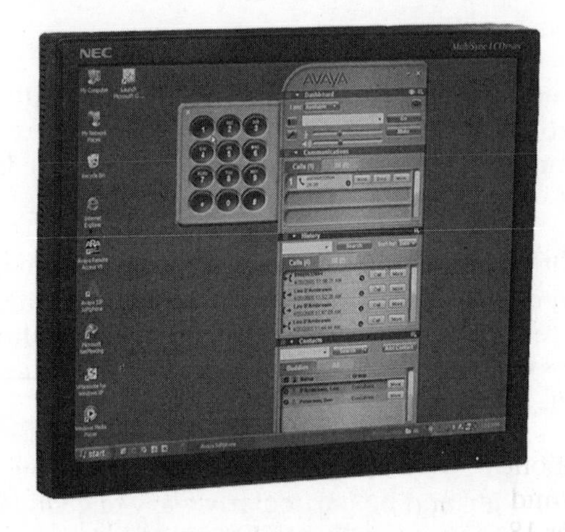

Figure 10-3: A soft phone running on a Windows PC.

Portable computers

Portable computers include laptops, notebooks, and tablet PCs. If you can carry it into a nearby coffee shop and carry a cup of coffee at the same time, it is a portable computer. One disadvantage of a portable computer compared to a stationary computer is the size of the screen for the dialing pad.

Another disadvantage is the cost of the computer itself. Here is a case where less is more. Portable computers are almost twice as pricy as desktop computers.

An advantage of using a portable computer is that you can take the computer with you, wherever you go. Today, portable computers are coming off the line as light as two pounds or less. My tablet PC weighs in at 2.2 pounds, but it cost almost $2000.

Portable computers are also more likely to include wireless networking capabilities.

Features supported

Soft phones generally don't provide the same level of calling features that you get with a hard phone. Soft phones run on your computer, and all features are implemented and accessed through the screen. Hard phones, on the other hand, include buttons and software optimized for quick use. The types of features provided with soft phones are generally limited when compared to hard phones.

The better versions of soft phones are generally Windows-based, so the software has a graphical user interface (GUI) that enables your computer to do VoIP telephony over the network. In its most basic form, the software needs to display the dialpad for making calls. It also needs to interface with your network using TCP/IP protocols.

Because soft phones work with Windows, you can use Microsoft Outlook and a Web browser to access contact lists (including LDAP-based directories) and the phone numbers stored within these applications. You can also do instant messaging and VoIP calls simultaneously. At present, this is about as good as it gets for call features.

Many versions of soft phone software are available. Most companies go with a proven market leader and standardize on a soft phone version for all employees. (See Chapter 18 for a list of top VoIP manufacturers.)

Consumers, on the other hand, can choose from (and may be bewildered by) a number of different flavors of soft phone software available over the Internet. See whether someone you know has installed the software and has had a good experience with it. When you look at which VoIP carrier to use, request a trial download of their soft phone software. If they require you to go and get your own soft phone package and merely sell you the VoIP carrier service, consider using a different carrier.

VoIP Wireless Phones

Several types of wireless phones are available. The first type are IP wireless phones, which have a limited range and are strictly tied to corporate networks. For example, a hospital or a large construction site may have wireless networking and VoIP available over that network. VoIP wireless phones hook into the network and do VoIP within their specified range. Features on these types of phones are generally limited.

One thing to watch for in IP wireless phones is whether they are WiSIP compatible. If they are, the phones can include quite a few features not normally available, such as the ability to connect to WiFi networks and IP-PBXs without the no-peak or off-peak minute charges. These types of phones cost a bit more, but they make calling other WiSIP phones very easy.

Finally, it can be argued that a pocket PC with VoIP capability is, indeed, a wireless phone. These types of computers do everything that a cell phone can do. If the pocket PC has built-in WiFi capability, you can use it to make VoIP calls in addition to regular cell calls.

Maximizing Your Current Telephone Investment

If you are pondering the move to VoIP but are thinking about your company's significant investment in existing telephone sets, read on. Your company is just like the majority of companies today still running on a PBX model, with lots of digital phone stations that were not cheap and a system that represents a sizable capital investment.

Upgrading older telephone systems

Most KTS and PBX models use digital telephones that have a great deal of capability and flexibility. (Both KTS and PBX systems are discussed in Chapter 11.) If your company is currently running a PBX manufactured in the last three years, it's a safe bet that all your telephones can operate in the new VoIP environment.

Some upgrade adjustments will be needed to your current PBX systems, but the individual telephone sets that connect to the PBX need no such upgrade. For instance, the PBX may need to have an interface card installed so it can connect to the LAN, but the individual phone sets would not need such an upgrade.

All devices in a VoIP network must have a MAC address, which is procured by installing a network interface card. Pure VoIP phones have their own MAC address, but when a PBX is being upgraded to work with VoIP, a NIC must be added to support this requirement. After the PBX gets its NIC and its own MAC address, all the telephone sets connected to the PBX can share the PBX's MAC address.

Non-VoIP digital phones use the inside house wiring to connect to the circuitry of the PBX. Nothing much changes on these phones except that the PBX that they have always connected to is now also connected to the VoIP network. Figure 10-4 illustrates this arrangement.

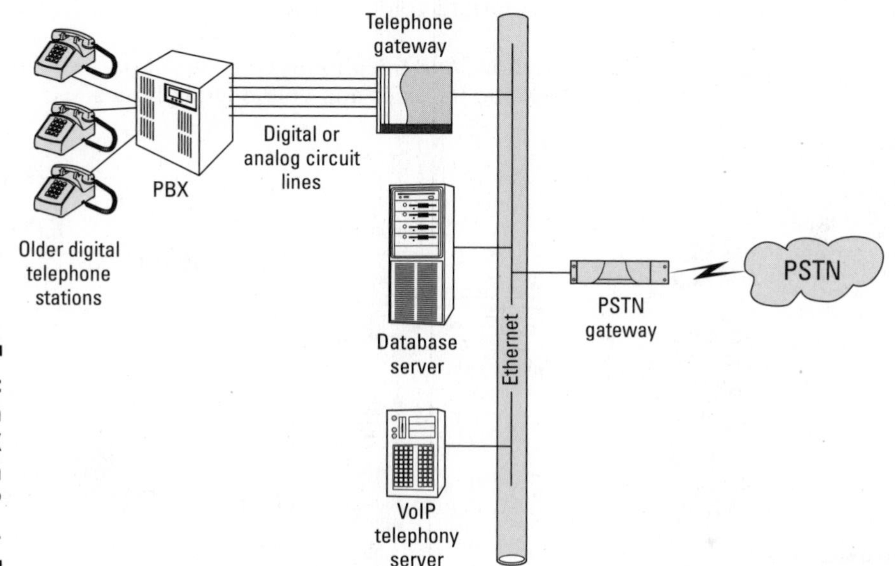

Figure 10-4: Using an older PBX system with a VoIP network.

These older telephones can enjoy the traditional features that come with any telephony connection. For example, voice mail can continue to be delivered via the network to the PBX telephone stations connected to it. Voice mail would be stored in the mailboxes that have already been allocated by the PBX administrator. Your company would not even have to reassign telephone numbers or mailboxes. Other features such as call transfer, call forwarding, and conference calling are all still available.

Using older telephones on the new VoIP network

One of the big differences between older digital phones and the newer VoIP phones is their respective call feature sets. For PBX telephones, traditional call features are delivered over the existing telephone wiring, not the LAN wiring or WiFi network utilized by the VoIP network.

However, these PBX telephones do not have a LAN connection port or the advanced features provided through VoIP telephones. If you need to provide any VoIP advanced features to a subset of your employees, your company must acquire the appropriate VoIP phones.

Figure 10-5 illustrates how VoIP telephones can coexist on a VoIP network with a traditional PBX and its older non-VoIP digital phones.

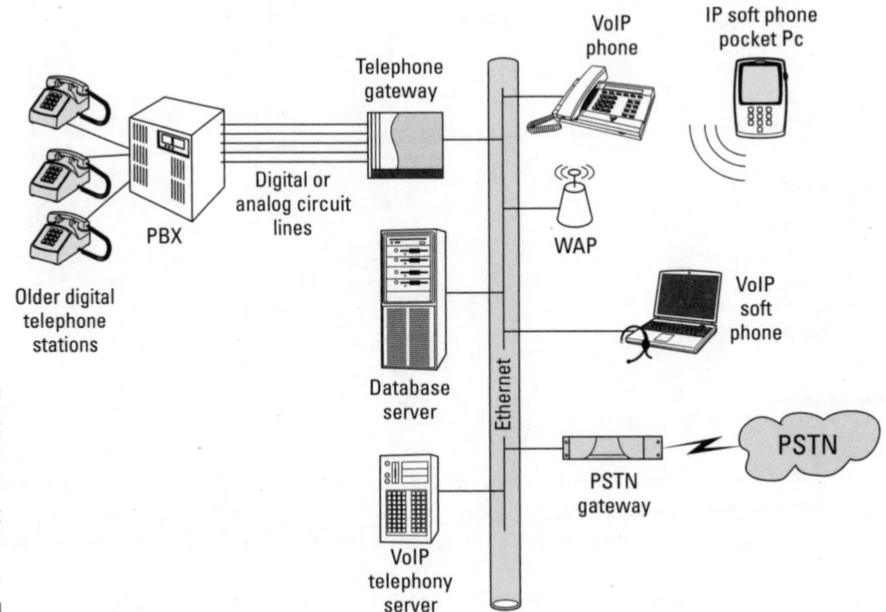

Figure 10-5: Using a PBX and VoIP equipment at the same location.

Part III
Making the Move to VoIP

In this part . . .

The bottom line is, well, the bottom line! The chapters in this part disclose exactly how you can figure out the bottom line for any VoIP conversion for your company.

Using base information and illustrative case studies, you discover how to compare costs and calculate savings. The numbers can tell you exactly whether a change to VoIP makes sense for you.

Making the move to VoIP can be challenging, regardless of how many locations your company has. The information in this part is essential to helping management focus on the benefits of any proposed conversion.

Chapter 11

Simplifying Cost Management

In This Chapter

- Watching the charges disappear
- Recapping the final four telephony models
- Unifying telephone and computer networks
- Convincing upper management to switch to VoIP

When VoIP was introduced, many analysts projected savings for companies choosing VoIP versus companies continuing to operate with circuit-switched telephony systems. But a number of these first-adopters ended up frustrated with the earliest forms of VoIP, mostly because the systems were based on using the Internet as their network backbone. (As you discover in Chapter 9, the Internet doesn't provide the optimal infrastructure for companies with heavy telecommunications needs.) As a result, the majority of companies back then continued to operate with circuit-switched telephony systems.

VoIP has matured since its inception, and today saves companies huge amounts of money. It provides a great quality of service over private, dedicated networks. Moreover, VoIP enables a number of slick calling options. VoIP features enhance the collaboration of employees across the enterprise and ultimately increase productivity while reducing the operating expenses of the company.

This chapter includes details that you need to consider when planning for a VoIP conversion. Here you find information about your current telephony system and how you can realize savings by converting that system to VoIP. You also find ways in which you can convince your corporate decision-makers that they should give the switch a try.

VoIP Comes and the Charges Go

One of the big "aha moments" with VoIP is that companies can enjoy an immediate cost benefit with their toll charges. VoIP can save money in other ways, as well. You won't need to pay any additional per-line feature charges because VoIP runs on your computer network. Regulatory fees, surcharges, and taxes are applied on a per-line basis. As you reduce the number of lines, the line cost and recurring charges go with them.

Reducing or eliminating phone lines

If you can eliminate one or more of the lines that you lease from the carrier, the call feature charges, the regulatory fees, and the taxes are also eliminated. Reducing the total number of lines really makes a cost difference. In the POTS/PSTN way of doing telephony, you get additional lines as you need to increase your capacity. In the VoIP way, you can upgrade your bandwidth on your dedicated line to increase capacity.

There are several disadvantages for companies that use POTS/PSTN rather than VoIP telephony. POTS-related lines are leased from a carrier. Just leasing a single line incurs added expenses. For example:

- Each line usually has a nominal startup charge.
- Each line has a monthly recurring access charge.
- For each POTS line, the company must pay monthly recurring usage charges. (Chapter 3 provides a breakdown of charges and charging categories.)

All recurring charges are based on a rate per minute per line. When you add up all the minutes from every line in operation at each of your company's locations, the monthly cost can get into serious amounts of money.

If your company has separate telephony and computer networks and the company has significant aggregate toll volume, you can reduce or eliminate most of your charges by converting to VoIP and running your telephony over your computer network.

Take off your add-on charges

Traditional phone service normally includes costs that apply to every single line you lease. Just like any other service, traditional telephony lines and services are taxed. Depending on where you are located, you could have one or more taxes in addition to all the other monthly charges. Taxes are based on the total cost of your line access and other services. For instance, for each line's total service cost, you can add the following taxes to the bill:

- Federal tax (about 4 percent)
- State tax (varies by state but the average range is 5 to 7 percent)
- 911 emergency surcharge fund (flat rate of $1 per line)

Taxes obviously affect your bottom line. VoIP, however, does not come with any taxes or surcharges. VoIP is totally unregulated and operates over your existing computer network. Therefore, taxes do not apply to your monthly bill.

Yippee! Deregulating your telephone costs

POTS/PSTN lines and services also involve other monthly regulatory fees. These are charges that go to various government entities. These fees are based on a percentage of each line's monthly access cost:

- Universal service fund (3.5 percent)
- Interstate access surcharge (20.9 percent)
- Telecommunications relay surcharge (0.1 percent)

These charges are based on a percentage of the monthly per-line access cost, but before you draw any conclusions about these costs being nominal, add up the number of lines and the total cost. Depending on where all of your locations are located (that is, which LATAs), these regulated fees vary somewhat. For a corporate customer, if you calculate about 4 to 7 percent of your total monthly access costs, you can get a close estimate. If you are a consumer, these add-on fees can be as high as 20 percent of your total monthly telephone bill.

With VoIP, you pay regulatory fees for your dedicated network transports, but you already pay these in support of your computer data network. VoIP runs over your packetized computer network, so you have no more taxes, add-on costs, or other regulatory costs for VoIP telephony.

Free call features

Calling features include items such as voice mail, call forwarding, call transfer, return call, and three-way calling. Traditional telephony requires you to pay a monthly charge for call features. These add-on charges may not apply equally to all the lines you lease because the features are optional.

Some call features are so popular, many people think they are a part of the telephone service and are expected to come with the access line. Voice mail, for instance, is considered an essential with any telephone, but you still have to pay the carrier $7 to $9 per month per line. If you use the popular call return feature (*69), you can pay around $1.00 to $1.50 for each use.

You can reduce the total cost of call features by setting up a bundled plan with the carrier. However, you do not have to add any call features to any line; they are truly options like a moon roof or climate control in an automobile.

VoIP comes with the usual call features that you have to either bundle with your traditional lines or pay à la carte per line as you use these features. But with VoIP, you don't need to worry about the cost of call features; they are all included at no extra cost.

Most companies use an internal telephone system, which can usually provide most if not all call features. However, with POTS and Centrex line models (see Chapter 2 and the next section), call feature costs are highly relevant to the company's monthly telephony bill. If your company has hundreds or thousands of lines, the overall cost for all features for all lines can be astronomical.

The Final Four Meet VoIP

To reduce the recurring charges for POTS telephony services, a company with fifteen or more employees, who each need a telephone, can acquire its own telephone system. Over the years, several conventional systems have emerged. All use POTS, but each one reduces the dependence on POTS lines and POTS line equivalencies. Also, they all provide traditional features (voice mail, conference calling, call transfer, and so on) at no extra cost. As a result, companies seeking to use conventional POTS services generally use one of the four non-VoIP telephony systems models. Larger companies may use one or more of these models, depending on the number of employees at each location. These models, which I call the final four, were introduced in Chapter 2.

Table 11-1 provides an overview of the final four, and the following sections describe each model in more detail.

Table 11-1 The Final Four

Model	*Equipment Location*	*Cost Structure*	*Comments*
POTS	Carrier lines run to company-owned phones.	Monthly recurring charges (MRC) per line, per phone. Regulatory fees apply to access line costs.	Call features are paid per month, per feature. Relatively high costs on a per employee basis. Not well suited for VoIP conversion unless toll charge savings justify conversion costs.

Model	Equipment Location	Cost Structure	Comments
Centrex	POTS-equivalent carrier lines run to customer's telephone on a per phone basis.	Higher POTS-equivalent line charges, and monthly recurring charges per line. Regulatory fees apply to access costs.	Little or no maintenance costs. Higher priced lines compared to POTS. Suitable for VoIP if company has substantial MRCs for either regional, intrastate, interstate, or international toll line carrier services.
KTS	POTS carrier lines run to the company's KTS switch.	Monthly recurring charges per line. Higher startup costs for KTS and phones. Regulatory fees apply to access lines.	Most features included at no extra cost (savings due to one POTS line for every six to eight phones). Suitable for VoIP if company has substantial MRCs for either regional, intrastate, interstate, or international toll carrier services (see Chapter 3).
PBX	Dedicated carrier transport lines to PBX. Bandwidth can be channelized to support VoIP migration.	Dedicated access lines. Highest monthly recurring charges. Bandwidth can be used for data, voice, and video. Can be used along with station phones in a VoIP network. Regulatory fees apply to access lines.	Call features available free. Call center capabilities. Higher monthly maintenance charges but reduced aggregate line costs. Avoids need for forklift upgrade because PBX can be integrated with VoIP. Highly suitable for VoIP because company carrier costs for regional, intrastate, interstate, or international tolls are reduced.

Goodbye POTS, hello VoIP

Does your company have fewer than fifteen phone stations or fax machines? Is it not bothered by high toll charges and requires no significant international services? If so, your company can stay with the POTS model. With POTS, the company does not even need to consider a different system of managing their telephony services; everything depends on the carrier. Figure 11-1 illustrates the traditional POTS model.

The company would work out an acceptable service plan with their carrier. They would need to have or acquire POTS-type telephones for each user. In the plan, the company should minimize what the per-minute usage rates are going to be for all regulated services categories. In addition, the company would establish what add-on charges are acceptable in the way of call features on one or more of the lines.

Because all employees will probably have voice mail, the company can cut some costs by getting voice mail for only a subset of the total number of lines to be leased. Voice mail usually comes with the ability to set up five or so mail boxes per line at no additional cost. If you get voice mail on the main number, you can set up a "hunt group" that rolls to the next line when the receptionist is unavailable, or you can have the caller leave a voice mail to any employee by leaving directions on the voice mail narrative (for example, "press 1 for Ms. Smith, press 2 for Mr. Williams"). Other call features can be added on the lines that need them.

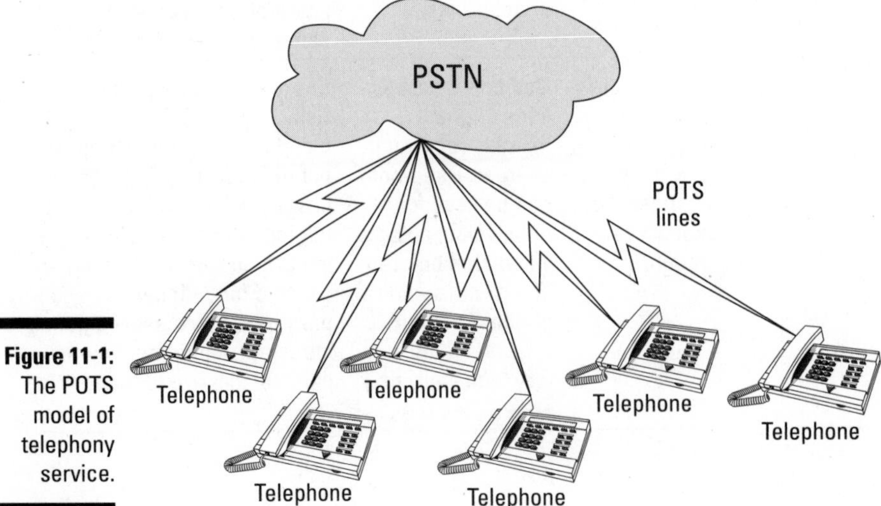

Figure 11-1: The POTS model of telephony service.

The bottom line with POTS is that each employee has a traditional phone. Some have more than one line to support a fax or dial-up modem services. The company gets the bill each month for all these lines, their respective service charges, and all add-on charges.

If your company has a computer network, you owe it to your financial health to at least take a look at VoIP. To get a total picture of what it is costing you or your company, divide your total monthly carrier services bill by the number of lines you are leasing for each employee. To isolate the total costs by line, usage, and cost per add-on item (taxes, add-on features, regulatory fees, and so on), create a spreadsheet in which each employee's line requirements reflect the monthly costs for that employee. Run the items per employee line horizontally across your spreadsheet. When you have gone through all of the billings, total the fields by column for each item. This gives you the total monthly cost breakdown for the lines and items you are tracking.

An example makes this easier to see. I'll show the costs for a small company with seven employees. The costs of their POTS lines and additional features is shown in Figure 11-2. Regulatory fees are outlined in Figure 11-3, and local and toll costs are shown in Figure 11-4. Their total MRC is $1644.

Figure 11-2: Tracking the costs of POTS lines and add-on features.

	Line	Line MRC	Voice mail	Call forwarding	Conference call	Caller ID	Call trace	Total
Main number	6200	$45	$7	$3	$0	$4	$4	$63
President	6201	$45	$7	$3	$4	$4	$4	$67
Secretary	6202	$45	$7	$3	$0	$4	$0	$59
Senior account rep (a)	6203	$45	$7	$3	$4	$4	$0	$62
Account rep (b)	6204	$45	$0	$3	$0	$4	$0	$52
Account rep (c)	6205	$45	$0	$3	$0	$4	$0	$52
Account rep (d)	6206	$45	$0	$3	$0	$4	$0	$52
Customer service rep	6207	$45	$7	$3	$4	$4	$4	$66
Main fax	6208	$45						$45
Modem (a)	6209	$45						$45
Secretary fax	6210	$45						$45
Modem (b)	6211	$45						$45
Modem (c)	6212	$45						$45
TOTAL		$585	$35	$24	$11	$30	$13	$698

	Line	Surcharges	Fees	Taxes	Total
Main number	6200	$15	$5	$6	$27
President	6201	$16	$6	$6	$28
Secretary	6202	$14	$5	$6	$25
Senior account rep (a)	6203	$15	$5	$6	$26
Account rep (b)	6204	$13	$4	$5	$22
Account rep (c)	6205	$13	$4	$5	$22
Account rep (d)	6206	$13	$4	$5	$22
Customer service rep	6207	$16	$6	$6	$28
Main fax	6208	$11	$4	$4	$19
Modem (a)	6209	$11	$4	$4	$19
Secretary Fax	6210	$11	$4	$4	$19
Modem (b)	6211	$11	$4	$4	$19
Modem (c)	6212	$11	$4	$4	$19
TOTAL		$171	$59	$66	$296

Figure 11-3: Tracking the costs of regulatory fees.

	Line	Local	Intralata	Instate	Interstate	Int'l	Total
Main number	6200	$3	$0	$0	$0	$0	$3
President	6201	$11	$39	$3	$7	$0	$60
Secretary	6202	$13	$4	$0	$0	$0	$17
Senior account rep (a)	6203	$8	$32	$1	$3	$0	$44
Account rep (b)	6204	$8	$36	$12	$4	$0	$60
Account rep (c)	6205	$8	$34	$11	$6	$0	$59
Account rep (d)	6206	$8	$42	$15	$4	$0	$69
Customer service rep	6207	$22	$65	$21	$6	$0	$114
Main fax	6208	$37	$76	$18	$2	$0	$133
Modem (a)	6209	$5	$0	$0	$0	$0	$5
Secretary fax	6210	$21	$45	$7	$2	$0	$75
Modem (b)	6211	$4	$0	$0	$0	$0	$4
Modem (c)	6212	$8	$0	$0	$0	$0	$8
TOTAL		$156	$373	$88	$34	$0	$651

Figure 11-4: Tracking the costs of local and toll calls.

Goodbye Centrex, hello VoIP-Centrex

The Centrex (Central Exchange) services model is owned and operated by the carrier. Centrex is physically the same as POTS; the line between the carrier company and your premise telephone service is the same. What is different is how and where the line terminates at the carrier end.

The lines run from the carrier's switching equipment to each customer's telephone. Figure 11-5 illustrates this model.

The number of lines needed for Centrex is almost secondary to the fact that the customer can terminate service at any time without penalty. Before VoIP, Centrex was a great solution for startups or companies unsure of their strategic plans because they could gain all the usual features along with POTS-equivalent telephony service quickly under a month-to-month plan. When the company's plans become concrete, they could terminate Centrex and convert to a new telephony system.

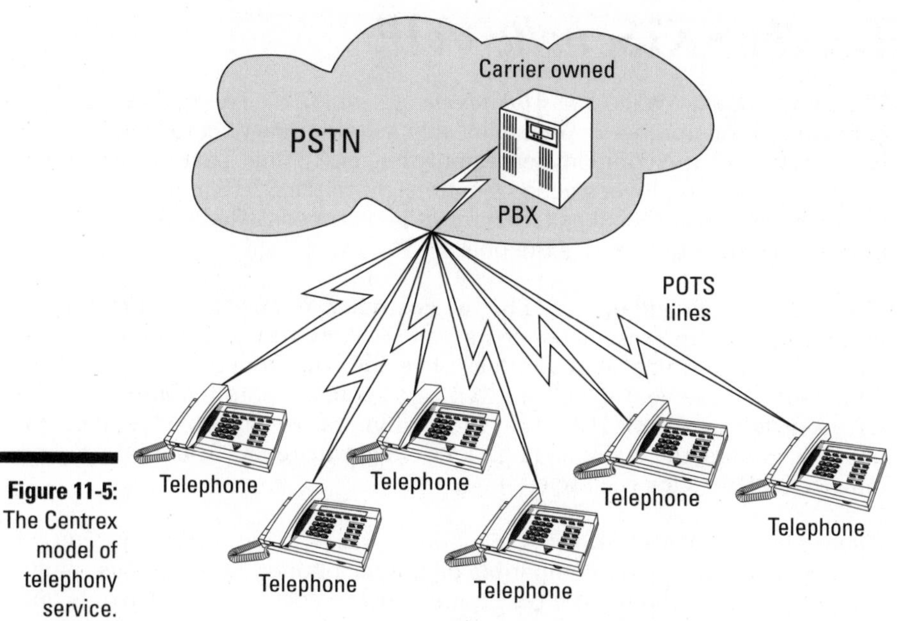

Figure 11-5: The Centrex model of telephony service.

Centrex costs more per month and per line, when compared to POTS, but can include many calling features without additional charges. Centrex can deliver services to the customer's telephone that are otherwise available only to users connected to more expensive high-end telephone systems.

VoIP can reduce or eliminate the need for POTS lines used for Centrex. It also provides calling features at no cost, so it can be good to switch from Centrex. VoIP gives you all the benefits of Centrex (and more) without the high costs. If you have a computer network and you like the idea of Centrex, consider VoIP Centrex. The process is practically the same as the traditional form of Centrex, except your company's voice signals travel over your computer network in packets on their way to your VoIP provider; the provider, in turn, sends them on to their destination.

To get a total picture of what Centrex is costing you or your company, divide your total monthly carrier services bill by the number of Centrex lines you are leasing. To isolate the total costs or costs by line item, you can follow the same procedure covered in the preceding section. Because the cost per line under Centrex is more than it is under POTS, you'll find that the regulatory fees you pay are higher also.

Goodbye KTS, hello VoIP

The third model is called a *key telephone system* (KTS). The KTS reduces a company's total number of phone lines. In fact, for every six to eight employees, on average, the company leases only one POTS line. That means a company with sixty employees needs a mere eight to ten POTS lines. The KTS provides many traditional call features at no extra cost. Reduction in lines means cost reduction across the board.

The KTS is owned and operated by the customer. You use the same physical lines, with the same associated costs, as POTS or Centrex. The difference is where the line terminates at your end. The lines run from the carrier to your KTS. You then use your own inside wiring to connect your telephones to the KTS. A disadvantage of KTS is that the customer is responsible for all maintenance (including the inside wiring), for configuring the KTS, and for programming call features for each telephone.

Figure 11-6 illustrates how the KTS model works. With some KTS systems, you may need to acquire compatible digital telephones; you can't just plug in any old analog phone. As a result, much of the cost of a KTS depends on the number of phones your company needs.

KTS was a good solution for smaller companies requiring more than fifteen lines. Now, however, KTS users can benefit by switching to VoIP because it gets rid of most if not all lines and uses the computer network for on-net voice traffic. If all of your company's locations are on-net, you not only reduce the number of lines required but also eliminate toll charges. And, in most companies, toll charges are the largest monthly telephony expense.

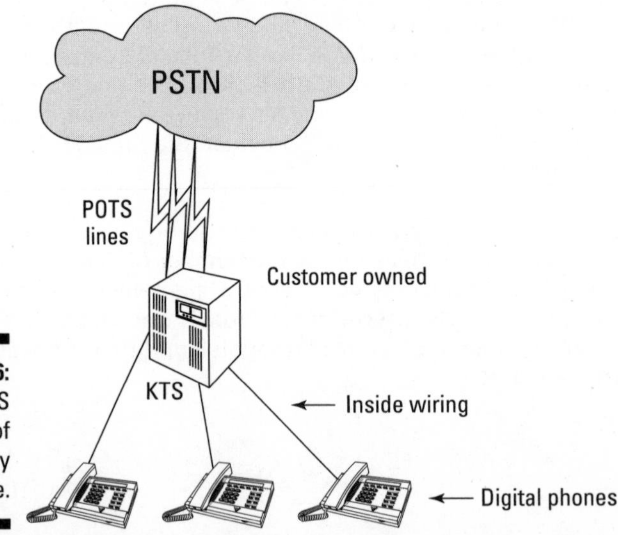

Figure 11-6: The KTS model of telephony service.

Goodbye PBX, hello VoIP-PBX

The fourth and final model of the final four is PBX, which stands for *private branch exchange.* Before VoIP, PBX was the mainframe of corporate telephony. KTS is a small version of PBX.

The most expensive of the four non-VoIP models, PBXs deliver the most value:

- PBX can use dedicated high-bandwidth lines out to the carrier or to other locations on the company's network.
- Interfaces can be used on PBX to provide full-motion videoconferencing.
- PBX has extensive call-management capabilities and the capacity for setting up and controlling multiple call centers.
- PBX can usually be upgraded to operate with VoIP. As a result, you can save money because you do not have to get rid of your PBX to go to VoIP.

As with KTS, companies using PBX can reduce the total number of POTS lines required by a factor of one line for every six to eight employees. But unlike KTS, PBX has the capacity for unlimited expansion. The largest workable KTS is limited to about sixty POTS lines; with PBX, you can have thousands of lines. Figure 11-7 illustrates the PBX model.

The PBX system's circuitry integrates multiple users over fewer lines at a single location and can also connect to all other locations.

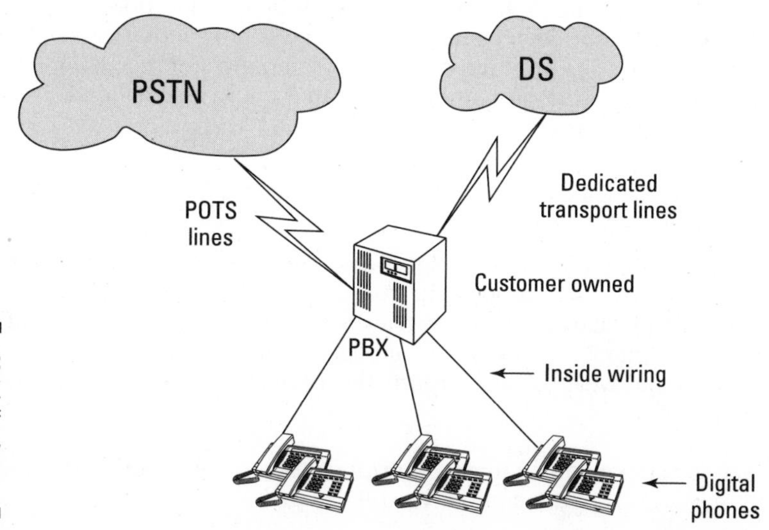

Figure 11-7: The PBX model of telephony service.

If your company has multiple locations, with each site having its own PBX, you can connect them all using a separate network, but the cost of maintaining a separate network is huge. If you have a great deal of recurring charges, the multilocation design for connecting all your PBXs can save your company big time. The recurring costs have to be leveraged against the costs of an entirely separate network. (VoIP runs over your computer network and does not require a separate PBX network.)

The PBX model provides great savings when compared to the other models, but it doesn't give you anything close to the savings attainable through VoIP. VoIP all but eliminates services charges for all on-net calls. VoIP reduces carrier services charges significantly for calls that travel off-net. For many large multilocation companies, these charges alone amount to millions of dollars per month.

Unified Networks

The new term for combining a company's telephony and data networks is *convergence.* A *converged network* is formed when all your data — including VoIP — travels over a single infrastructure. Another popular term for this is *integrated networking.* Integrated networks incorporate the use of computer data, telephony signals, and video signals onto the same network.

VoIP networking is about unifying the people that work for the company. VoIP, among many other attributes, is a unifier.

VoIP is not just about reduction of lines and carrier services charges, although these can be significant. It also adds features and functions that the company never dreamed of. As a result, VoIP enhances productivity. Use of the unified network increases productivity for the company and its customers. The following are two examples among many that could be provided.

Larry's story

Larry is a Human Resources Specialist who works for a kitchenware manufacturer. His office is located at the company's headquarters in southwestern Pennsylvania. Larry frequently travels among the company's twenty-three locations, which include plants and sales offices spread out across the Midwest and West coast. He routinely conducts interviews with new employee candidates.

Larry's company used to pay enormous toll charges, but VoIP eliminated a whopping 92% of the company's toll charges.

Every time Larry traveled to a different company location, the IT staff would set him up with a computer network connection and a telephone. Now that his company has switched to VoIP, Larry merely has to plug in his computer, which runs IP soft phone, at any available port. He doesn't need to have anyone from IT make special configuration changes for him. He can even use IP soft phone to reroute his extension to any phone in the office.

Moreover, with his new IP soft phone, Larry can make use of videoconferencing from his laptop computer. He no longer needs to travel to other sites to conduct screening interviews. The candidates report to the company location nearest them, and the sponsoring location allows the interview candidate to use one of their VoIP phones that run videoconferencing. Through VoIP, Larry's company saves in both toll charges *and* travel costs.

Joann's story

Saving on travel and maintenance costs are only part of the picture. VoIP truly enhances productivity, delivering feature-rich applications that let you do your work and still have a life.

Consider Joann, who works for one of the top health insurance providers headquartered in the Northeast. Her company has seventeen locations connected over a VoIP network. Throughout a typical day, Joann uses a VoIP-enabled telephone to receive announcements, make phone calls, and send and receive e-mail.

Joann starts her day by checking her IP telephone's Web page for announcements. One morning, she read that her friend and coworker Rae Lynn had a baby boy the night before. She made a note to send Rae Lynn's family a card.

As part of her job, Joann reviews healthcare claims that do not fit the normal criteria for a final decision by the utilization review (UR) department. Much of Joann's communications relate to the status of the claims she is investigating. She regularly communicates with people located at her home office and other sites. Joann also interacts with staff from her company's huge healthcare provider network to determine the details of each claim she receives for disposition.

With the exception of any calls made in the local calling area, all Joann's telephone calls are carried on the corporate VoIP network. When the call is to a provider located off-net near one of the company's other locations, the call travels from Joann's VoIP telephone over the corporate VoIP network to the distant site's location, where it goes over the company's LAN at that location, out the gateway, and into the local calling area. As a result, for all Joann's telephone calling, her resulting monthly off-net charges are minimal.

All the claims Joann's company receives are transmitted to their UR department through the Web. If a claim cannot be approved for payment upon receipt, the UR department forwards it electronically on the corporate VoIP network to Joann, with a copy to the medical director of the respective source location and a copy to the headquarters' medical director.

Joann works frequently with the medical director at the headquarters' location because of the technical nature of many of the claims. On average, Joann calls this medical director seven to ten times per day on claim-related matters. Therefore, she includes this medical director in her VoIP telephone buddy group and makes full use of the *presence* feature indicator on her VoIP telephone. If the presence indicator is lit, she knows not to waste her time calling the medical director because he is on the telephone with someone else. Joann also has a presence indicator set up for people she works closely with.

Much of Joann's day is spent on her VoIP telephone. She uses it to process inbound or outbound e-mail from the company's various locations. Sometimes the content of a claim requires Joann to contact other personnel in the company. When she needs to do this, Joann accesses her browser-based directory to retrieve the person's contact information and automatically dials the VoIP telephone number with one click. Or if Joann is on the road, she can simply speak the name of anyone in the directory, and the VoIP speech-access application dials automatically.

Needless to say, Joann is a busy woman. About thirty minutes before her workday ends, she checks the weather advisory corner of the Web page on her VoIP telephone. She wants to know whether she needs to bring her umbrella when she heads over to the subway station. She checks her voice mail and typically opts to have the remaining unheard messages printed so she can read them on the ride home.

Convincing Your Boss

Part of gaining the support for the move to VoIP convergence is to convince the company that it is the right move. You generally need sign-off by the people who manage the company's technology and, to a certain extent, the staff that reports to these managers. But you know and I know that by and large getting approval comes down to convincing upper management.

The best way to appeal to upper management is to focus on the cost-effectiveness of convergence. Map out your current expenses and contrast those numbers with the expense of VoIP convergence. The numbers speak for themselves: VoIP convergence reduces operating expenses enough to pay for itself in the near term, and it can save the company a whole lot of money going forward. (For examples of how other companies have accomplished it, see Chapters 12 through 14.)

Another benefit that can speak to upper-level managers is that implementing a unified, integrated network brings the company together, and makes all employees reachable on a higher, horizontal plane of communication. It promotes collaboration, enhances productivity, and ultimately leads to an increase in revenue.

Last, you need to provide your management with a seamless plan for transitioning to the new system, as discussed in the following section.

A seamless transition

The good news is that introducing VoIP onto your computer network can be performed while keeping your conventional POTS/PSTN telephony systems operational. Because the two are physically separate networks, they can operate simultaneously.

If you work with a carrier company that supports VoIP-based telephony, and a hardware vendor that provides hardware to support both types of networks, you can enjoy your conversion to VoIP while still having the safety net provided by the older system.

Providers typically offer reduced costs to keep the old systems running while you install the new VoIP-based systems. When you are comfortable with your new converged and integrated network, you can plan for the removal of the old telephony systems and the termination of any non-used carrier services.

If your company has made a significant investment in telephony systems that were not IP-ready in the last few years and those systems are capable of being upgraded to handle IP telephony, your company can now plan the move to VoIP while still protecting the company's investment in older systems. This is another less costly way to reap the full benefits of your original systems while positioning your company for the eventual full conversion to VoIP.

Whatever VoIP conversion option your company may choose, you will be running a single network that integrates computer data with voice. (Video conferencing can be added too.) The requirements for managing the company's network become more unified versus divided. A single comprehensive network management system can be used to monitor and control everything traveling over the network. Fault-isolation can be more readily implemented because you don't need to troubleshoot what network the problem may be on.

Because your company will unify its support staff into one department, the ensuing cross-training and convergence experience gained by all in this department can result in a reduction of the company's dependence on outside experts. In the short-term, your company may need to use outside contractors while using existing providers. In this way, needs are optimally supported until the conversion is at or near completion.

Meeting your future with VoIP

In a competitive marketplace, companies that are forward-thinking look at their competitors. Market projections based on a mere percentage of the total telephony marketplace indicate that the VoIP telephony market could grow to as much as $15 billion a year by 2008.

This trend means one or more of your competitors are making the move to VoIP and enjoying the benefits. It also means that your company will be at a disadvantage if it doesn't undertake a strategic plan to convert. As collaborative companies with a unified workforce satisfy their customers in unprecedented ways, they are going to increase their respective market share. Your company may not be able to afford to ignore VoIP technologies much longer.

Bandwidth on demand

Besides the movement of the market, including your competitors, toward VoIP, you need to evaluate a few significant technical benefits. First, VoIP networks support the kinds of transport services that run packetized services not only for computer data, but also telephony voice and video where needed. These transports are usually dedicated lines of substantial bandwidth capacity.

As discussed in Chapter 7, bandwidth is normally *channelizable,* which means that the bandwidth of the line can be divided into channels. The channels can be used *dynamically* (whenever they are needed for a specific application that is seeking to run on them at any point in time). When channels are not needed, they go back into a pool of channels for other applications, including data, voice, and video needs. This type of operation is often referred to as *bandwidth on demand.*

To achieve this type of bandwidth usage, the network architecture uses terminating equipment called *level three switches.* Providers that supply the transports usually include or specify exactly what model of switch fits the bill. Bandwidth on demand is a function of network design that works very well with VoIP.

Scalability, as needed

Scalability refers to the degree to which your company can make changes to support growth and increase access to and use of the network. On the data side of the network, Ethernet is highly scalable. New users, IP telephones, computers, and other devices can be connected to the network on a plug-and-play basis.

When an employee needs to move to a new location in the building, for example, his or her IP telephone and computer can be unplugged and taken to the new location, where they are plugged back in. Both devices relearn automatically on startup the identity of the employee, so the devices are operational immediately. Consider the benefit of this capability: No one needs to go to the telecommunications closet and reprogram port numbers or change network addressing information.

VoIP protocols bring a certain degree of intelligence to the enterprise network that makes change a pleasure and a joy rather than a frustrating and time-consuming hassle.

Costs to move, add, and change

As a result of the high degree of scalability and the intuitive intelligence of VoIP networks, move, add, and change (MAC) costs are a thing of the past. In companies that still operate under one or more of the traditional telephony models, expensive MAC costs are incurred whenever an employee moves, they need to add new users, or they need to make changes in a user's telephony profile.

In traditional telephony models, one of the major annual expenses is the maintenance of the telephone system. MAC costs are the single most expensive item on this budget line. In traditional networks, MAC costs average from 17 to 31 percent of the entire maintenance budget. Under a VoIP model, you have no need for a MAC cost item in your budget.

Older telephony system technicians that complete these MAC changes bill at $150 per hour. Imagine if the company had to make a major set of moves or changes. For example, if a company needs to move a department from the fifth floor to the twelfth floor, traditional MAC costs can run an average of $90 per employee. When new inside wiring is required or the telephony system needs to be reprogrammed, the costs are even higher. Under VoIP, everyone unplugs on the fifth floor, goes to their newly assigned space on the twelfth floor, and plugs back in — no outside intervention is needed! Under VoIP, MAC changes go away. Again, more cost savings and time savings to justify your move to VoIP.

Chapter 12

Locations Galore

In This Chapter

- Considering VoIP over multiple locations
- Evaluating your current networks
- Creating a plan
- Implementing the plan
- Reviewing the plan's success
- Savings and the bottom line

Businesses with multiple locations are prime candidates for benefiting from VoIP, but they also face challenges that never beset consumers or single-location companies. If you are running a business with multiple locations, this chapter is for you. Here you find guidance and examples of how you can implement VoIP with a minimum of hassle.

Challenges of Multiple Locations

Companies that have worked with and mastered traditional networks (either telephony or data) often think that it's just a small step to implement VoIP. Think again! If your business has multiple locations, don't even consider implementing VoIP on your own. You'll want to work with a qualified VoIP partner. Working with a partner well-versed in VoIP can help save you money, and the money you save from the conversion more than pays for the partner's services. Your company can then gain the skills to gradually become experts in VoIP the same way they did with the KTS or PBX telephony systems — a little bit at a time.

Many people, without realizing it, develop a POTS mentality, thinking of communications problems in terms of old line-based solutions. To adapt to the VoIP model, you need to shed this old-style mentality and look for ways you can effectively converge your data and telecommunications networks.

If you can see that the POTS side of what your company needs is the "small potatoes" part of a VoIP project, congratulations! Your challenge now is to recognize that you don't start any project — including VoIP — by focusing on the small potatoes. It's much more important to figure out how you integrate VoIP into your multilocation computer network.

When you move to VoIP, you are putting your telephony systems onto your computer network. With all the new features VoIP brings to the table, it seems like you're multiplying your telephony applications exponentially, similar to how computer applications seem to multiply on your data network. (I won't ask how many computer apps you have because I know it's numerous and grows and churns every week.)

Don't try to apply a traditional telephony design model to the implementation of VoIP. The companies that do this end up failing or not doing as well as they could have. You need to adapt the traditional telephony models when you go to VoIP.

Evaluating Your Existing Networks

If the VoIP partner you select can't do a thorough analysis of your existing networks, including all monthly billings, you need to make an executive decision and select a partner that knows how to do it. The result of the network evaluation should be a complete spreadsheet that lists each company location and all one-time and recurring charges that your company is paying.

It's amazing how many companies don't have a clue about what they are paying. In their defense, evaluating the costs can be complicated if the company has multiple carriers and a plethora of calling services and in-house telephony systems. I've had clients who had multiple PBXs, with a Centrex line for every employee in the company — even though the value of Centrex is that it eliminates the need for the company to go to the expense of having its own in-house telephone systems such as PBX.

When you do your analysis, start with your monthly billings. The billings paint the picture of your costs.

To illustrate what I mean, consider the case of a medical practice I helped convert to VoIP. Bremer Healthcare (a fictitious name) has eleven sites in southwestern Pennsylvania and lots of local and intralata traffic. Each site has its own LAN. The WAN setup to connect them all was far from perfect. They initially called me to help them connect to their health insurance companies over the Web.

Bremer was told that they needed to start submitting their billing claims over the Web within six months, or they could lose a good part of their business. This led to completely reevaluating their entire telephony and computer network infrastructure. They didn't even have a dedicated connection to the Internet — they were running slow dial-up connections into a popular ISP.

Figure 12-1 shows the results of my initial cost analysis of their telephony and computer network systems.

No doubt you were hit with some shock and awe when you saw the huge and unnecessary charges for Bremer's intralata services. The number of POTS lines they used was also unnecessarily large. The results of the cost analysis and the fact that they have multiple locations made them primary candidates for VoIP.

Bremer Healthcare started their medical offices in outlying suburban areas. Because the sites were spread out across all regional toll areas, they were getting clobbered on charges, paying anywhere from $.10 to $.42 per minute. They paid their phone bills each month as a necessary cost of their business, not knowing that they could save huge amounts of money.

Bremer's costs didn't end with their traditional telephony charges; they also paid large fees for dedicated carrier services. The shame of the situation was that although Bremer had a dedicated computer data network, the only application running was e-mail. Four of the locations could not even connect to the e-mail server at the main site. They had no high-speed Internet access. Figure 12-2 shows the cost analysis for their dedicated services.

Existing circuit-switched costs

Location	Local calling area	Voice lines	FAX lines	Modem lines	Total lines	MRC access	Local	Intra-LATA	Intra-state	Inter-state	Int'l	Monthly usage	MRC total	Annual
Main - Carnegie	Pittsburgh	27	2	14	43	$3,139	$5,134	$21,664	$58	$21	$0	$26,877	$30,016	$360,192
Brentwood	Pittsburgh	3	1	1	5	$325	$712	$4,793	$0	$0	$0	$5,505	$5,830	$69,960
Bethel Park	South Hills	4	1	1	6	$390	$685	$4,065	$0	$0	$0	$4,750	$5,140	$61,680
Canonsburg	Washington	4	1	1	6	$390	$918	$4,781	$0	$0	$0	$5,699	$6,089	$73,068
McMurray	Washington	4	1	1	6	$390	$548	$5,664	$0	$0	$0	$6,212	$6,602	$79,224
Robinson	Moon	4	1	1	6	$390	$429	$2,465	$0	$0	$0	$2,894	$3,284	$39,408
Upper Saint Clair	South Hills	4	1	1	6	$390	$756	$1,132	$0	$0	$0	$1,888	$2,278	$27,336
Mount Lebanon	Pittsburgh	4	1	1	6	$390	$682	$2,832	$0	$0	$0	$3,514	$3,904	$46,848
Cranberry	North	3	1	1	5	$325	$721	$10,093	$11	$0	$0	$10,825	$11,150	$133,800
AlleghenyValley	North	3	1	1	5	$325	$537	$9,114	$16	$0	$0	$9,667	$9,992	$119,904
Wexford	North	3	1	1	5	$325	$759	$11,328	$10	$0	$0	$12,097	$12,422	$149,064
TOTAL		63	12	24	99	$6,435	$11,881	$77,931	$143	$21	$0	$89,976	$96,411	$1,156,932

Figure 12-1: Telephony service cost analysis.

Existing packet-switched costs				
Location	Local calling area	T1 lines	MRC access	Annual
Main - Carnegie	Pittsburgh	2	$982	$11,784
Brentwood	Pittsburgh	1	$519	$6,228
Bethel Park	South Hills	1	$575	$6,900
Canonsburg	Washington	1	$725	$8,700
McMurray	Washington	1	$675	$8,100
Robinson	Moon	1	$575	$6,900
Upper Saint Clair	South Hills	1	$675	$8,100
Mount Lebanon	Pittsburgh	1	$525	$6,300
Cranberry	North	1	$505	$6,060
AlleghenyValley	North	1	$522	$6,264
Wexford	North	1	$958	$11,496
TOTAL		12	$7,236	$86,832

Figure 12-2: Dedicated services cost analysis.

Bremer Healthcare was told by their carrier that because of the way their frame-relay network had to be designed, two T1 lines were needed at the main location. This proved to be false. I used one of the two T1 lines to give the new VoIP network a dedicated Internet gateway and recommended that they drop the other one as an unnecessary expense.

Developing a Plan

Because Bremer Healthcare's existing telephone system was centered on the circuit-switched network and their existing dedicated network was not used for any critical applications, there was room to install VoIP without bringing down their day-to-day business functions. The analysis of their existing telephone and computer network infrastructure made it obvious that a hub-and-spoke design using VoIP would resolve most if not all of their problems.

Designing a VoIP solution

The VoIP plan for Bremer Healthcare required more use of dedicated lines, which they ironically had in place and were already paying for. Greater usage of the dedicated transports reduced the total number of POTS lines needed. The plan called for the use of a PSTN gateway at each site, providing connectivity to the PSTN. The PRI 23-channel transport line at each site was used to anchor each LAN to the PSTN gateway. This action consolidated off-net local calls through the PSTN and substantially reduced their need for POTS lines.

Using the PSTN gateway with PRI lines ensured that Bremer would be able to maintain connectivity to the PSTN as they reduced their dependence on POTS lines. Essentially, one PRI at $325 per month equals twenty-three POTS lines. They were paying $65 per month per POTS line. If you do the math, not counting their horrific intralata usage costs and local calling usage, they were paying $6435 per month just for the POTS lines. The cost for eleven PRIs, one per site, cost them only $3600 monthly (11 x $325). It's a no-brainer!

The plan provided on-net telephony calling, which eliminated their horrendous intralata calling charges. It provided a dedicated Internet gateway, a state-of-the-art firewall, and an intranet at the main site. All the other LANs have access to this intranet as well as the Internet through the LAN at the main location. Even with the increased functionality, Bremer's overall monthly operating expenses were substantially reduced.

The new VoIP network integrated telephony systems and computer data systems onto the same infrastructure. As a result, the company raised their capabilities to a never-before-seen level, all while lowering their overall costs.

Putting your plan into action

To implement the VoIP plan, Bremer Healthcare needed six essential components:

- Ethernet network interface card (NIC)
- VoIP gateway
- VoIP server
- VoIP-compatible switches
- VoIP voice mail
- Internet firewall

The first component was an Ethernet network interface card in each telephone system (PBX). With their existing telephony systems connected to the Ethernet, they could make use of all their existing digital telephone stations. This does not in itself provide each user with all the new and exciting features available through VoIP, but it maintains the level of telephony services they had before VoIP, at far less expense to the company. It also sets the stage for adding VoIP-enabled telephones.

The second component was a VoIP gateway for each of the LANs in the company. The main site would receive a high-level VoIP gateway because it needed to terminate multiple T1 lines and support other enterprisewide functions, such as the Internet gateway for all the LANs. All the other LANs would receive a much less expensive version of this VoIP gateway. The VoIP gateway devices would provide the main interface between each LAN and the newly redesigned WAN.

The third component was a properly configured VoIP server at each site. High-capability servers were chosen with maximized clock speed, lots of memory, and more hard-disk capacity than the average server. This allowed the VoIP servers to back each other up in case one should go down. Should a problem develop, the network could be used to swap in a new server and download the complete VoIP image configuration from one of the remaining servers.

The fourth component was to replace all LAN switches running at each of the locations with VoIP-compatible switches. This component did not need to be implemented immediately, but eventually they would want VoIP telephones; adding the switch upfront was easy because we would need to bring the LANs down for a period of time anyway. So, it made sense to implement this component upfront, particularly because the VoIP cost savings were so dramatic.

The fifth component pertained to storing voice mail. Management installed a new voice mail server, but did not enable it throughout the company during the first year. Instead, they continued to use voice mail primarily on the old telephone systems. However, Bremer also started testing and implementing new VoIP phones. As a result, five VoIP-enabled phones were requested for managers at the main site and three at each of the other LAN sites. These phones were to use the new VoIP mail server for voice mail.

Finally, the sixth component was the placement of the Internet firewall at Bremer's main site. It became the main ingredient in support of security on their newly defined intranet and their gateway to the Internet. All employees from all the other LANs could use the intranet for Internet-related applications (browsing, e-mail, chat, and so on). Those fortunate few who received VoIP phones were even able to set the Web browser function on their phone's screen (see Chapter 10).

Obviously, these six components have costs associated with them, but many of these costs are one-time startup costs. Figure 12-3 shows the details of Bremer Healthcare's VoIP startup costs, taking into account the six components they needed.

VoIP startup costs

Location	Local calling area	Upgrade existing PBX/KTS	VoIP partner consulting	VoIP servers	VoIP gateways	VoIP switches	Voice mail server	VoIP phones	Internet gateway firewall	Total
Main - Carnegie	Pittsburgh	$7,000	$98,226	$25,750	$47,000	$34,000	$21,000	$2,400	$31,000	$266,376
Brentwood	Pittsburgh	$2,475	$9,044	$14,000	$16,000	$9,000	$0	$800	$0	$51,319
Bethel Park	South Hills	$3,140	$9,458	$14,000	$16,000	$9,000	$0	$800	$0	$52,398
Canonsburg	Washington	$2,817	$9,281	$14,000	$16,000	$9,000	$0	$800	$0	$51,898
McMurray	Washington	$3,224	$9,572	$14,000	$16,000	$9,000	$0	$800	$0	$52,596
Robinson	Moon	$3,100	$9,575	$14,000	$16,000	$9,000	$0	$800	$0	$52,475
Upper Saint Clair	South Hills	$3,640	$10,155	$14,000	$16,000	$9,000	$0	$800	$0	$53,595
Mount Lebanon	Pittsburgh	$2,890	$9,325	$14,000	$16,000	$9,000	$0	$800	$0	$52,015
Cranberry	North	$3,245	$9,685	$14,000	$16,000	$9,000	$0	$800	$0	$52,730
AlleghenyValley	North	$3,000	$9,496	$14,000	$16,000	$9,000	$0	$800	$0	$52,296
Wexford	North	$3,450	$9,784	$14,000	$16,000	$9,000	$0	$800	$0	$53,034
TOTAL		$37,981	$193,601	$165,750	$207,000	$124,000	$21,000	$10,400	$31,000	$790,732

Figure 12-3: Startup costs for Bremer.

Staging the Implementation

When implementing a VoIP system, the major consideration is network downtime. Whenever it is necessary to swap out major components such as switches and servers, taking the network down is a consideration.

In the case of Bremer Healthcare, downtime wasn't a big concern because they chose to maintain their existing POTS-based telephony systems. Their data network did not currently support any of the company's strategic objectives. This provided the flexibility to move all components in without a corresponding problem of having to take either the telephony systems or the computer network systems down for long periods of time.

In Bremer's case, installation was scheduled for weekends, with one weekend per location. The first task was to install the VoIP server and the Internet firewall at the main site, and then install the components site-by-site using Internet connectivity as a test of success. The full testing of the VoIP system would be performed as each site came up and then again after all sites were up.

Plug-and-play

Ethernet LANs provide a plug-and-play environment. This means you can often plug devices into the network, and they learn what other devices are connected and how to work with them. Plug-and-play makes it easy to add devices (including VoIP devices) to the network. However, not all devices or applications are as easy to add; some require quite a bit more work to place on your LAN.

For example, if you were to install a Network Management System (NMS), you would have to set parameters and configure it to run the way you want it to run. Bremer Healthcare did not use a separate NMS, choosing instead to incorporate NMS functions into the firewall installed at the main site.

The implementation of VoIP at Bremer required the installation of many new Ethernet devices. For instance, at each Bremer site, we swapped out the older non-VoIP switches with plug-and-play VoIP switches. We installed PSTN gateways, connecting each to the PRI line installed by the local carrier. All that was necessary was to plug the line into the proper port on the gateway and then program the settings. Lastly, the gateways used to terminate each LAN's T1 connection were installed to enable connectivity to the other Bremer locations. Installing the gateways wasn't quite as easy as plug-and-play because each required some configuration to properly recognize the network and the T1 line and to enable security settings.

Managing downtime

In any VoIP conversion, there is downtime, and the best way to manage it is to anticipate when it could happen and then plan for it. Unfortunately, downtime always seems to be the result of factors you can't control.

The major downtime with Bremer Healthcare's conversion was coordinating with the local exchange carrier to switch over the T1 lines. The carrier was apparently not happy with the new design that Bremer planned. As part of the conversion plan, Bremer had provided the carrier with a spreadsheet of scheduled dates, times, and locations for the changes to the dedicated T1 lines. Basically, the conversion called for keeping the existing T1 lines but changing them from a frame-relay configuration to a dedicated, private T1 service.

Coordinating this change was the biggest headache in the conversion. Only two of Bremer's eleven locations were converted on schedule. Fortunately, Bremer was not using the frame-relay service for much of anything anyway and didn't have a problem with the downtime that occurred.

Reviewing the Effect

Your IT folks can tell you technically how well your converged network is running and how to keep it humming, but only your workforce and your customer base can tell you if it is truly working. More companies are going to VoIP than ever before because of two primary benefits: increased features and reduced expenses.

Features and costs of the new VoIP network

In the case of Bremer Healthcare, all the players — doctors, clinical staff, administrative staff, patients, and even suppliers — were overjoyed with the results of the VoIP conversion. The new network had an immediate effect on monthly expenses, and the numerous call features inherent to VoIP promoted enhanced productivity, employee mobility, and new options for communications. Figure 12-4 shows the revised cost structure for Bremer's circuit-switched network after the VoIP conversion.

Figure 12-4: Circuit-switched costs after VoIP.

VoIP circuit-switched costs

Location	Local calling area	POTS lines	FAX lines	Modem lines	Total lines	Access lines	PRI	MRC access	Maintenance services	Local	Intra-LATA	Intra-state	Inter-state	Int'l	MRC usage	MRC total	Annual
Main - Carnegie	Pittsburgh	4	2	3	9	$585	$325	$910	$545	$3,185	$24	$21	$34	$0	$3,264	$4,719	$56,628
Brentwood	Pittsburgh	2	1	1	4	$260	$325	$585	$125	$645	$2	$0	$0	$0	$647	$1,357	$16,284
Bethel Park	South Hills	2	1	1	4	$260	$325	$585	$100	$472	$1	$0	$0	$0	$473	$1,158	$13,896
Canonsburg	Washington	2	1	1	4	$260	$325	$585	$125	$698	$2	$0	$0	$0	$700	$1,410	$16,920
McMurray	Washington	2	1	1	4	$260	$325	$585	$81	$412	$4	$0	$0	$0	$416	$1,082	$12,987
Robinson	Moon	2	1	1	4	$260	$325	$585	$81	$405	$2	$0	$0	$0	$407	$1,073	$12,879
Upper Saint Clair	South Hills	2	1	1	4	$260	$325	$585	$125	$635	$3	$0	$0	$0	$638	$1,348	$16,176
Mount Lebanon	Pittsburgh	2	1	1	4	$260	$325	$585	$125	$570	$4	$0	$0	$0	$574	$1,284	$15,408
Cranberry	North	2	1	1	4	$260	$325	$585	$140	$684	$46	$6	$0	$0	$736	$1,461	$17,527
AlleghenyValley	North	2	1	1	4	$260	$325	$585	$125	$540	$37	$8	$0	$0	$585	$1,295	$15,540
Wexford	North	2	1	1	4	$260	$325	$585	$140	$725	$41	$4	$0	$0	$770	$1,495	$17,935
TOTAL		24	12	13	49	$3,185	$3,575	$6,760	$1,712	$8,971	$166	$39	$34	$0	$9,210	$17,682	$212,180

If you compare Figure 12-4 to Figure 12-1, you'll notice the sharp reduction in the number of POTS lines required: from ninety-nine down to forty-nine. The addition of the PRI transport lines, to maximize bandwidth for off-net calls, helped to consolidate much of the old POTS configuration.

Maintenance services were also added to the picture, which provided a safety net that Bremer didn't have before. Lastly, the big-ticket item that was virtually done away with was the costly intralata expenses. With the VoIP network, most of the intralata traffic was put on-net, thus bypassing the metered expenses of the PSTN.

It's not just a new way to do circuit-switched

The circuit-switched aspects of the VoIP network are only half the picture. To connect all the sites and enable them to use the Internet, a private dedicated infrastructure was needed. The plan called for converting Bremer's underutilized frame-relay network to a dedicated private network. Figure 12-5 details the postconversion costs associated with the packet-switched side of the network.

VoIP packet-switched costs

Location	Local calling area	T1 lines	MRC access	Internet access	Annual cost
Main - Carnegie	Pittsburgh	1	$575	$825	$16,800
Brentwood	Pittsburgh	1	$519	$0	$6,228
Bethel Park	South Hills	1	$575	$0	$6,900
Canonsburg	Washington	1	$725	$0	$8,700
McMurray	Washington	1	$675	$0	$8,100
Robinson	Moon	1	$575	$0	$6,900
Upper Saint Clair	South Hills	1	$675	$0	$8,100
Mount Lebanon	Pittsburgh	1	$525	$0	$6,300
Cranberry	North	1	$505	$0	$6,060
AlleghenyValley	North	1	$522	$0	$6,264
Wexford	North	1	$958	$0	$11,496
TOTAL		11	$6,829	$825	$82,773

Figure 12-5: Dedicated services costs after VoIP.

If you compare Figure 12-5 to the preconversion amounts in Figure 12-2, you notice that the costs didn't go down much. Most savings were due to the VoIP design eliminating one of the two T1 lines at the Bremer main site.

The big story with the dedicated side of the network is that the carrier was taking advantage of Bremer. They used to have twelve T1 lines running frame-relay services, and four of their sites couldn't even connect. After the conversion, Bremer had eleven T1 lines with full, dedicated bandwidth available to them — and the overall costs didn't go up. Moreover, all eleven sites had Web access through the main site's Internet gateway.

Bottom-Line Analysis

In the end, it is the cost of business that determines if you can continue to *do* business. The good news is that VoIP helps you continue business like never before. If you're having trouble justifying the productivity-enhancing features of VoIP (increased mobility, agility, and customer satisfaction), look at the bottom-line financial analysis.

For example, Figure 12-6 provides an executive summary of before-and-after costs related to Bremer's communications and data networking.

Annualized financial analysis				
		Circuit-switched	Packet-switched	
Item	Description	networking	networking	
1	Recurring cost of existing system	$1,156,932	$86,832	
2	Recurring cost of new system	$212,180	$82,773	
3	Annualized savings	$944,752	$4,059	
4	Gross annual savings			$948,811
5	VoIP startup costs			$790,732
6	Net annual savings (year 1)			$158,079
7	Payback period			10 months

Figure 12-6: Summary of Bremer Healthcare VoIP savings.

Before their VoIP conversion, Bremer had no one in-house who could manage their communications, and only one young employee to fix computer problems. With the savings from VoIP, Bremer was able to hire several technicians so they could develop the in-house network expertise they needed.

As Bremer's IT manager stated, "VoIP's value is priceless." What they gained was much more than what can be expressed in dollars and cents, even though the dollars and cents are impressive. Bremer's immediate savings paid back their startup costs in approximately ten months — faster than they had expected. Even with startup costs, Bremer saved $158,000 in their first year of using VoIP. This provided an unadjusted projected return on investment (ROI) of $948,000 for the second year.

Chapter 13

Setting Up the Smaller Office

In This Chapter

- Finding out whether VoIP will work for you
- Reading the small print
- Evaluating your current networks
- Saving with VoIP
- Summing up the savings

Considering VoIP for a single-site, smaller company has a lot of parallels to putting VoIP in at a larger multilocation company (see Chapter 12). There are also some big differences. These differences typically pertain to the company's mission and scope, the volume and categories of calls, and the company's strategic plans for the future. It may not seem like these three things are critical for a smaller company, but they determine if VoIP is a good fit.

This chapter focuses on the needs of smaller companies, particularly those with single locations. You'll find out about a company that may closely parallel your own and see exactly how they benefited from switching to VoIP.

Is VoIP for You?

If a single-location company does the majority of their business within a three-mile radius of their location, the majority of their calls are likely in the local calling area. In this case, there would not be enough of a reduction in toll charges to offset the startup costs for VoIP. VoIP is not for everyone; in this case, I would help the company make the most cost-effective use of their existing POTS infrastructure.

On the other hand, if a single-location company does the majority of their telephony business outside the local calling area, we need to take a look at the monthly expenses for telephony and computer data networking.

Many single-location companies fit this situation. For example, I consulted with a company that ran six call centers out of a single location. This company does more telephony business in a week than some of the Fortune 500 companies do in a month. Their monthly invoices for phone services were in the multiple millions of dollars. They called everywhere in the United States and the world and received calls from everywhere in the world for their many clients. Such a company actually redefines the meaning of a small, single-location company.

The analysis in this type of context would pretty much follow that of Bremer Healthcare, the case study in Chapter 12, with only a few variances. Instead of location line items for each site (as in Bremer's case), an analysis would use call center line items and calculate the traffic patterns and costs. My experience with call centers is that they usually have competent people at the helm of their telephony services, and if they are not running VoIP, they have the best of the best that you can derive from traditional telephony services. These folks tell the carrier what they want and jump all over them when they don't get it. As a result, I'm more concerned with all the other single-site, smaller companies that directly depend on carriers. These are the companies that get taken to the cleaners more often than not.

Location alone does not give you all the information you need to recommend a conversion to VoIP. To make sense in a small company, VoIP has to change the cost picture.

In small, single-site companies with largely local calling area telephony usage, the risk is that VoIP could become an added expense that may not have a payback and may not change the company's productivity. These are crucial points that a smaller company needs to think through before making a VoIP decision.

The bottom line is that VoIP has to either reduce your operating expenses or help in some way to increase your revenue.

Figuring out those contracts

After a thorough analysis of how your single-location company makes and receives its telephone calls, you can determine whether a move to VoIP is worth making. That said, you also need to know your company's long-term plan so that you can factor your goals into any VoIP design.

Another significant factor is whether or not the company is encumbered by any current contracts. If they just signed a three-year lease for non-VoIP PBX and all its telephone stations, the lease costs must be considered in any plan. Any carrier services contracts can usually be terminated without penalty or can be modified with a new term, but now running with VoIP.

After current monthly billings, the long-term plan, and any outstanding contracts are figured into the potential for going with VoIP, a complete analysis can be performed to financially support whatever decision is made.

Current costs meet long-term plans

A client of mine, Keystone Mortgage (a fictitious name), had a long-term mission that included increasing its presence both regionally and within the state of Pennsylvania. Ninety-five percent of their business was conducted over the phone. They were doing so well in their area that they expanded into the outlying regional areas. They were also looking at putting an office in Philadelphia because they were told that doing so would enable them to call customers in those areas more cheaply.

When I sat down with Keystone's owner and president, it was clear he wanted Internet access for all Keystone's agents. He was concerned because telephone bills had gone through the roof since they expanded their regional calling. He told me that he didn't understand why they had to pay so much to give everyone their own voice mail box or to do simple things like have a conference call. He also expressed dismay over the fact that for all Keystone was paying, they could not get better Internet services, e-mail for everyone, or a company Web site.

Five of the agents had dialup modem lines for the Internet, and they also paid $25 per month per line for unlimited access to the Internet. They were pondering two additional modem lines but thought the cost per month was high. Everyone had to work out times when they could share the computers tied to the modem lines. Big uploads and downloads were a problem because the dial-up connections were slow. No one enjoyed this arrangement.

Analyze bills and contracts

I told Keystone Mortgage's president that it appeared their telephony systems were inadequate and that there were better, more cost-effective alternatives. I said the alternatives could improve their telephony systems with features such as voice mail, the Internet, e-mail, and a Web site.

I told him that the best way to start was to analyze their current infrastructure. All I needed was a copy of all monthly billings related to telephone systems, computer systems, and any contracts Keystone might be obligated to at present.

Evaluating Existing Networks

After you analyze your company's billings and contracts, you should end up with a spreadsheet that lists all one-time and monthly recurring charges by line item. When your company sees what they are paying, it's amazing how quickly they lose any preconceptions that they are a small company. It is typical for clients — even small single-site companies — to not realize the total cost of their current infrastructure.

It became immediately apparent that Keystone Mortgage could benefit from a VoIP telephone system. They could pay for it with the savings they would realize from reducing POTS lines and toll charges. They had no dedicated transport lines in the picture. They had a LAN and were paying a monthly lease charge for an e-mail server, but the company leasing it required an additional contract to actually make the server work in Keystone's LAN environment. Keystone's president had not signed the contract for the extra work because one of the agents, who had some tech knowledge, said he could do the extra work at no cost, but it was not happening.

There were so many things that this company was doing in a less-than-optimal way; the situation begged for a line-item analysis, as shown in Figure 13-1.

Keystone's annualized expenses added up to nearly $156,000. They were pouring money down the drain. When I showed the president the spreadsheet in my follow-up meeting, he was overwhelmed and highly motivated to listen to solution alternatives.

Breaking down the costs of POTS telephony

Much of Keystone's recurring costs were wrapped up in their use of intralata calling. Each of the agents had a daily calling quota, and they reached their quota each and every day. The average number of calls per agent per day was thirty-eight to forty-seven, including twenty-one intralata calls. The agents averaged three intrastate calls per day, but they wanted to expand this even more. The remaining calls were to the local calling area. Keystone also had significant outbound fax calls that added minutes to all calling areas.

Keystone Mortgage had no idea of the carrier usage cost of the calls they made. They did not distinguish calls to the local area from calls to the local toll area. (See Chapter 3 for more on call types.) Most people draw a distinction when the call is across the state or even out of state. It was no surprise that the company's expansion into the nearby regional toll areas added horrific usage charges. This factor alone made them a candidate for VoIP.

Existing Tradtional Infrastructure (voice)														
			Circuit-switched											
			Access						Usage					
Location	POTS lines	FAX lines	Modem lines	Subtotal volume	POTS lines	MRC access	Local	Intra-lata	Intra-state	Inter-state	Int'l	MRC usage	MRC total	Annual
Upper Saint Clair office	15	2	5	----	22	$1,606								
Local calling	255	34	30	7018			$351							
Intralata calling	48510	6468	0	54978				$8,796						
Intrastate calling	7920	176	0	8096					$1,295					
Interstate calling	1320	44	0	1364						$55				
International calling	0	0	0	0							$0			
SUBTOTALS						$1,606	$351	$8,796	$1,295	$55	$0	$10,497	$12,103	$145,240
		Add-on call features												
Call trace	4.25												$64	
Voice mail	6.5												$98	
Call forwarding	5.25												$79	
Conference call	4.75												$71	
SUBTOTALS	20.75												$311	$3,735
		Add-on services												
						MRC								
Internet service provider			5		$25	$125							$125	
SUBTOTALS						$125							$125	$1,497
		Other charges (nontelephony)												
E-mail server lease													$418	
Ethernet switch 10/100Mbps													$34	
SUBTOTALS													$452	$5,424
TOTAL														$155,896

Figure 13-1: Keystone Mortgage's cost analysis.

Breaking down the costs of computer networking

There was even more to consider when it came to computer networking. To do VoIP correctly, you need a computer network somewhere in the picture. Keystone had a fledgling Ethernet LAN running off a leased LAN switch. They were paying an outside contractor $34 per month for the privilege of connecting their computers to this switch.

Keystone was paying the same contractor $418 per month for an e-mail server. The contractor installed the server and connected it to the LAN switch. All computers could see the server on their individual computers but could not connect because the employee computers did not have a required piece of software installed. The contractor delivered this software (called the client software) with the server, but the contract for the billable time to install the client software was not signed. In addition, no provisions had been made to connect either the LAN or the server to the outside Internet.

Putting VoIP to Work

The optimal solution to Keystone's problems could be realized by setting up an efficient data network capable of handling VoIP. The company needed to add a multifunction gateway to their LAN. To connect to the Internet, they needed a dedicated T1 line that would terminate on this gateway. The T1 line would provide the level of digital bandwidth needed for both data and VoIP.

They also needed an Internet service agreement to provide access and VoIP telephony hosting services over this dedicated T1 line. The agreement would provide e-mail accounts for all, a Web site for the company with unlimited storage, and their own domain name. Internet access over the company network would put an end to the need to share computers with dial-up lines. Keystone would now be able to do business and advertise their services on the Web at high bandwidth levels.

The existing lease contracts did not have a lock on them, so Keystone was able to terminate them immediately, resulting in a savings of $452 per month. The new VoIP servers more than replaced whatever applications this lease arrangement was trying to accomplish.

Too good to be true?

Any VoIP conversion has a few one-time equipment charges. For clients that get the willies over these charges, I offer them the option of leasing the items, inserting language into the contract that says they can yank it all out if it doesn't do the job. It's easy to understand how some companies get worried about startup charges, especially if they've been burned before.

For most customers, leeriness crops up when asked to shell out $18,000 or so for a VoIP gateway that presumably does everything from soup to nuts. Imagine the following conversation between a customer and the VoIP consultant:

Client: "So let me make sure I understand you. You're telling me that if I sign up for this service, this gateway thingamajig will let me carry all my on-net calls for free?"

Consultant: "Yes."

Client: "And I won't get much of a bill for any long-distance calling, right?"

Consultant: "Right!"

Client: "Which means I save close to $9000 a month, right?"

Consultant: "That's right."

Client: "That sounds too good to be true. Something has to be wrong with that picture."

Consultant: "I can give you references of other companies that did it. We can also arrange for your agreement to have language that permits you to go back to your old company's network at no cost should the results not meet expectations."

Just because something sounds too good to be true doesn't mean that it really isn't true. Many companies receive immediate savings from VoIP that really are amazing.

Supporting your telephony calls

For supporting calls primarily to the local calling areas, I recommended a PRI line. (There was no getting around Keystone's local calling requirements.) The majority of toll calls and faxes could be made "toll free" over the dedicated T1 line through their new hosted VoIP service. The savings would include calls or faxes to any toll area anywhere in the country and enable Keystone to expand their market reach to any area in the United States with no additional toll charges. Before VoIP, they were trying to figure out how to get the revenue to pay for a new office across the state in the Philadelphia region. Now they were happily looking at how to add agents to their present location, if and when needed, to sell their services across the country.

Much of the cost for the conversion to a VoIP environment would come from their savings on recurring toll charges. To understand how it is possible to manage these costs to your company's benefit, take a look at Keystone Mortgage's costs, as illustrated in Figure 13-2.

VoIP Infrastructure (voice & computer data)																
					Circuit-switched											
			Access								Usage					
Location	Local calling area	POTS lines	FAX lines	Modem lines	Subtotal calls monthly	Subtotal POTS lines	PRI line	MRC access	Local	Intra-lata	Intra-state	Inter-state	Int'l	MRC usage	MRC total	Annual
Upper Saint Clair office	Pittsburgh	2	2	2	----	6	$325	$763								
Local calls (nonmetered)		284	12	0	6512				$326							
Intralata call minutes		24	0	0	24					$4						
Intrastate call minutes		0	2	0	2						$0					
Interstate call minutes		0	0	0	0							$0				
SUBTOTALS								$763	$326	$4	$0	$0	$0	$329	$1,092	$13,109
			Add-on services													
								MRC								
VoIP Internet Service Provider								$2,495								
SUBTOTALS								$2,495							$2,495	$29,940
			Dedicated access													
							T1 line	MRC								
Upper Saint Clair office							$695	$695								
SUBTOTALS							$695	$695							$695	$8,340
TOTAL ANNUALIZED																$ 51,389
Start-up costs		Network equipment and consulting services														
	Upgrade existing PBX	VoIP servers (2)	VoIP gateway	VoIP switch	VoIP phones	VoIP consulting		Total								
TOTAL ONE-TIME COSTS	$2,745	$21,378	$18,477	$3,225	$7,400	$8,400		$61,625								

Figure 13-2: Cost analysis after VoIP conversion.

Understanding VoIP savings

For smaller, growing companies, it may be a bit difficult to grasp the real savings realized with a VoIP conversion if you look at only monthly charges. It is much more effective to consider annualized costs and savings. (In Figure 13-2, you can see the annualized savings on the right side.)

For Keystone, the first item that changed considerably is the sharp drop in recurring toll charges. The revised annualized figure for circuit-switched, POTS-related telephony is just over $13,000. Local calling is now supported over a 23-channel switched-access PRI line, and toll-related calling is now supported by a T1 line running VoIP services. POTS lines dropped from twenty-two to just six. These changes resulted in an annualized savings of more than $132,000 when compared to Keystone's previous method of making calls. That magnitude of savings provides the company with room to grow and change.

Next you'll notice that the call feature charges are gone. (See Chapter 3 for the dirt on traditional add-on charges.) In a VoIP system, all the usual call features and many new and exciting features come with the system for free. This is one advantage of using your computer network and the TCP/IP protocols to carry your telephony calls.

The next figure that sticks out is a whopping $2495 per month for dedicated VoIP and Internet access. Many would say that is a lot of money for Internet access. Note, however, that this amount is not for only traditional ISP service but also includes a few other critical services. These other services add tremendous value to the VoIP design for a single-site smaller company that has no full-time IT staff. You should take note of these added-value services before drawing any quick conclusions about lack of value:

- Configuration of gateway routing functions to support TCP/IP on the LAN
- Carrying unlimited VoIP telephony calls at no additional cost to anywhere in the country
- Secure firewall services
- E-mail accounts for all employees (with unlimited storage)
- Hosted Web site
- Domain name registration
- Support services 24 x 7

These added services solve a large part of Keystone's headaches and support their strategic mission plans to the maximum. This solution can occur only if Keystone uses the dedicated T1 line that runs from the company's gateway to the VoIP/Internet provider.

As you can see in Figure 13-2, Keystone's startup cost for the conversion was just under $62,000. That's a lot of cash for a small company to cough up, but now they are rolling in the dough. They got their money back in six months, revenues are up 28%, and they are talking about expanding into other states and renting the floor above them for new staff. Soon they will no longer be considered a small company.

Financial Analysis

For small single-site companies, demonstrating the cost benefits of VoIP is not overly difficult. Keystone Mortgage, for example, had only fifteen people in the company, but they had an annualized expenditure of $156,000 for their existing infrastructure. For that kind of money, the staff should not have to share dial-up modem lines for Internet access.

Expansion plans should not be complicated by unnecessary costs resulting from traditional circuit-switched phone systems. Many small single-site companies could probably identify with the savings experienced by the implementation of VoIP at Keystone Mortgage. Just compare what they were getting and paying under the old system and the new system.

More importantly, look beyond the money savings and see the other types of value they gained. They have become a much stronger company with all kinds of scalability. Morale has improved along with productivity. They were able to expand into markets across the state without having to move anything, add anything, change anything, or even pay anything more.

Figure 13-3 provides the financial analysis summary for the Keystone Mortgage case study. It totals their revised monthly operating expenses, savings, and payback period. This summarizes the savings for both the circuit-switched portion of their infrastructure and the dedicated, packet-switched portion.

Annualized financial analysis				
Item	Description	Circuit-switched networking	Dedicated packet-switched networking	
1	Annual recurring cost old system	$155,896	$0	
2	Annual recurring cost new VoIP system	$51,389	$8,340	
3	Annualized savings	$104,507	($8,340)	
4	Gross annual savings			$96,167
5	VoIP startup costs			$61,625
6	Net annual savings (year 1)			$34,542
7	Payback period			9 months
8	Value of new VoIP network going forward			Priceless

Figure 13-3: Executive summary of expenses after VoIP conversion.

When I first began working with Keystone Mortgage, the president had no idea what a T1 line was, and in fact did not know such a thing existed. He thought all telephone lines were the same and that all phones just plugged in and worked. He was totally blown away when he got his new VoIP phone and was able to have the company's new Web page on the phone's screen. He said, "Great, now I can surf to the stock ticker Web site and get the latest rates." He was starting to see immediate benefits of the VoIP conversion.

Chapter 14

Providing Dollars and Support

In This Chapter

- Figuring the costs of VoIP
- Investing in VoIP
- Understanding least cost routing
- Furnishing the support
- Keeping current

The costs of VoIP can be difficult to track. VoIP runs on your computer network; therefore, most of your costs are not calculated the same way traditional telephone services are.

If you are a consumer or a small business with fewer than fifteen employees and lots of toll charges each month, the cost savings with VoIP will easily offset your POTS telephone bill. If you are a company with many people using lots of minutes on the public telephone network each month, you will want to use your highlighter and calculator for this chapter. The same holds true if you are with a larger company that has multiple locations.

Evaluating VoIP Costs

Consider a case where you have two distant locations supported by traditional telephone systems, which means one or more PBXs (see Chapter 11). The general rule was that companies with more than a hundred employees would use the PBX model, putting at least one PBX at each location. With the PBX model, the customer typically owns, operates, and maintains the telephone switching system.

PBXs are capable of using dedicated access lines or switched access lines. But to connect locations over thousands of miles, the company would likely use dedicated T1 lines.

In a PBX model, the carrier is responsible for providing the outside access lines into each PBX, but the customer takes care of any hardware or software changes that must occur inside. A PBX is like having your own switchboard. Most of the costs associated with the PBX model are fixed. Normally, the external costs that cover what type of access lines you use are also fixed but billed monthly.

However, monthly recurring charges are not fixed. Depending on the services being used, these charges can add up to a huge monthly expense. Many larger companies (those with hundreds of locations) pay millions each month just for recurring carrier charges. VoIP is the technology that can reduce, if not eliminate, a large portion of these types of charges.

Before you can consider converting to VoIP, you need to have a handle on your existing costs. How else can you explain to your company's management why you want to change?

If you are in a new company just starting out or you have a branch location that is moving into a new building, the selection of VoIP to carry your telephony services is likely a no-brainer when it comes to the math. You still need to be diligent in researching and listing all your telecommunications costs, breaking them down according to one-time charges and recurring charges. You'll find that individual, recurring line items are overwhelmingly targeted to be either reduced or eliminated by VoIP.

Gathering cost data

Some costs associated with VoIP are similar to the voice services of old, but other costs will definitely change. You've discovered, in earlier chapters, both the differences and similarities between VoIP and traditional telephony. From a business perspective, it's beneficial to try to make sense of the differences and similarities in your particular situation by identifying the costs associated with VoIP.

VoIP hardware, depending on your configuration, can be purchased and depreciated like any capital investment, or it can be leased and addressed as an expense item. In any VoIP conversion, you also must take into account maintenance costs for both hardware and software.

To make a case for replacing an existing voice system with a VoIP system, you need to be able to show the current expenditures for voice services and compare these to the projected expenditures and savings of the new system.

At a minimum, you need to identify the following items as they pertain to whatever traditional system is being used at each location considering a change. Try to get three to six months of billings for each item:

- Line costs that currently support voice traffic
- Line costs that currently support data traffic
- Monthly costs for local, intralata, intrastate, interstate, and international
- Hardware costs associated with your PBX or KTS
- Digital telephone stations
- Maintenance costs (hardware, software, support services)
- Training and in-house expertise (PBX and KTS models)

These items help you get a good idea of your overall voice costs as they relate to existing systems. (Similar information was used in putting together the financial justifications presented in Chapters 12 and 13.)

To estimate your costs in a VoIP world, a different set of figures is required. Start with projections for the following information for each site considered for conversion:

- Transport line costs
- Monthly costs for local, intralata, intrastate, interstate, and international
- Hardware costs associated with upgrading the network to support VoIP (such as gateways and VoIP switches)
- IP phones (hard, soft, and wireless)
- Maintenance costs (hardware, software, support services)
- Training

By gathering the information listed in the preceding section, you are in a better position to approximate an apples-to-apples comparison.

Performing comparisons

VoIP challenges every preconception you may have had about telephony. In traditional telephony design models, component cost comparisons were less complex. They tended to follow a system hierarchy made up of various subsystem components, some of which had other subsystem components. For example, a PBX might have a backplane circuit chassis into which you could install circuit-pack subsystem boards to connect lines. Boards and lines had to correspond. A T1 line had to have a DS1 circuit back, and so on.

In a VoIP design model (also called a converging network model), the network becomes the system. The computer data network is now called the *foundation network*. Traditional telephony networks are converged onto computer networks to run not only data but also voice and video. Indeed, VoIP runs on and requires the computer network.

Managing your costs by line item and relating these costs to their product and services is the optimal way to illustrate and justify how a VoIP design can reduce your telephony costs. The nexus of savings from VoIP is determined by two important line item cost factors. One is the cost management of your network's transport lines. VoIP can use the same lines that may be in operation to connect your company's network. In some cases, no additional lines may be needed. The other factor is that VoIP can reduce your recurring usage charges for intralata, intrastate, interstate, and international telephony service categories. VoIP can also reduce the number of POTS lines required and significantly reduce your local calling recurring charges.

Transport lines

With VoIP, all toll-related voice traffic that previously traversed traditional phone lines between locations A and B now traverses the VoIP WAN line at no additional cost using the T1 transport line. In addition, each site can use a gateway to connect their LAN to the PSTN using a PRI line for local calls. The PRI replaces twenty-three POTS lines.

When calculating how much bandwidth is required to support VoIP traffic over the data network, remember the rule among telephony experts: one line or POTS line equivalent for every six to eight users. This ratio is open to modification, however, based on what percentage of their day each user spends on calls. (Your VoIP partner should be able to assist you with this calculation. If not, you may want to select another partner.)

After you have the number of POTS line equivalencies needed to support your VoIP traffic over the T1, multiply that number by either 25 (the bandwidth, in Kbps, needed for compressed traffic) or 80 (the bandwidth for uncompressed traffic). This gives you the amount of dedicated bandwidth, in Kbps, needed to support your VoIP traffic over the existing WAN data line. Armed with this information, you can contact your WAN circuit provider and obtain the cost of the additional bandwidth along with any additional supporting changes. This is one area where you will see a cost increase because you are adding additional bandwidth to an existing line, but these costs should be incremental.

Recurring usage charges

These costs for recurring usage can be a little tricky because calling patterns can change, so I always try to err on the conservative side. To project what your toll calls will cost you after conversion, look at the average of the last six months' worth of toll charges. Summarize all the calls made between locations A and B (the two sites you are converting to VoIP). Subtract the average total costs of these calls from your actual costs under VoIP to achieve your projected savings.

Ask your voice provider to give you your billing in a soft copy format, in addition to your regular printed bill. A soft copy is your monthly bill on a compact disk (CD), usually in Excel or Word format. This makes it easier for you to sort, merge, and summarize data to get it in the format you need.

Some VoIP components do not play well together. Do your homework and work closely with a trusted partner to lead you in the right direction. Although all systems may claim to support industry standards, most vendors add an extra layer or features that may not translate well between disparate systems.

After you have identified and settled on your vendor, figuring out hardware and software costs is relatively straightforward. A good place to start is to provide your wish list to your vendor and await the numbers.

When selecting hardware, don't fall into the cheapest component trap, especially when it comes to switches and handsets. Handsets can be powered in a number of different ways, and each handset has different power requirements. Manufacturers handle power differently as well. Going the cheap route, you may find yourself with a 24-port switch that is capable of powering only twenty VoIP handsets because the handset and switch do not negotiate power properly.

Another piece of essential hardware is the uninterruptible power supply (UPS). A UPS provides power in the event of an outage. Without this component, you lose the phone if you lose power. In the PBX model, the voice company includes a battery backup system in the event of a power outage, so you always have a dial tone. You need the same functionality for VoIP. You need backup power not only for your VoIP PBX and associated hardware, but also for your handsets. Make sure you purchase a UPS matched to your needs and that all VoIP components are plugged into the UPS.

Costs related to personnel

The complexity of VoIP systems is different from traditional voice systems in some respects. Traditional systems typically run proprietary operating systems and vendor-specific applications. Configuration can be performed by internal staff, after they are properly trained, or through a contractor. In a traditional voice system, the telephone company managed the end-to-end network. The voice network carried only voice traffic over its own dedicated infrastructure internal to the company or facility. Troubleshooting and problem resolution were addressed in this isolated network. The traditional voice network was also built for redundancy and fault tolerance.

With VoIP, voice traffic is carried over the same network as data traffic, using the same protocols as data traffic and sharing the same bandwidth. Many of the same personnel and systems used to manage data networks can be used to manage VoIP networks. Thus, the biggest difference between traditional phone systems and VoIP is in how they are managed and maintained.

In the traditional pre-VoIP system, the voice provider managed the entire external infrastructure up to the PBX (and in some cases including the PBX). VoIP systems can be managed the same way, or they may be implemented and supported internally. The components of the VoIP system require a different set of skills to implement and maintain.

Making the Investment

Investing in any new technology is about not only how you are going to pay for the technology, but also understanding what you expect to achieve from the technology. VoIP is no exception. If you set expectations early in your project and define key performance indicators by which you can measure success, your transition will go more smoothly.

Depending on how you plan to pay for the transition — buy and depreciate versus lease and expense — the costs may vary. Each approach has advantages and disadvantages and the choice to purchase outright or lease is based on many factors, including cash flow, money on hand, projected earnings, company culture, financial stability, and the life cycle of the technology.

If you purchase outright, you can depreciate the costs over a period of time, but there is an initial outlay of capital. You will probably have to keep the system for the full depreciation cycle, so be sure that cycle matches the technology life cycle. Otherwise, you may end up with outdated technology that is costly to upgrade. Maintenance costs also escalate as the system ages.

The lease-and-expense approach allows you to stretch out payments over the life of the lease. You'll want to make sure your lease fits your term needs and be especially careful of how the lease ends. At the end of the lease term, there will probably be a buyout, which can be anything from a $1 buyout to a fair market value buyout. The fair market value is usually set by the leasing company and may be more then you anticipated. With the leasing approach, make sure you have someone experienced in the art of leasing involved early in your cost-evaluation exercise.

You can negotiate the best deal by knowing what you need, the different products, and how much you are willing to spend. This may sound like common sense, but common sense often goes out the window when it comes to unfamiliar technology. You need to know your business needs, tailor your system requirements to these, and understand what you are purchasing. You also have to know how the system is maintained and who does the maintaining.

Cost-Effective VoIP Designs

As you find out in Chapters 12 and 13, implementing VoIP at a single location doesn't always provide as big a return on investment as using VoIP to connect multiple locations. In a single-site installation, your VoIP system functions like the traditional PBX. The real cost savings are realized when you connect locations and use your wide area network (WAN) to carry voice and data between locations.

Least cost routing can also offset some of your long-distance charges. Here is how *least cost routing* works: Suppose you have two locations, one in San Diego and one in Pittsburgh. You connect these two locations with a dedicated transport and eliminate any toll charges between the locations. With least-cost routing, if you want to call Los Angeles from Pittsburgh, the VoIP PBX can be configured to route your call through the network before it finally goes off-net in San Diego. The cost for the call is lower because it costs less to call from San Diego to Los Angeles than it does from Pittsburgh to Los Angeles.

Simple, right? Not so fast. How are you going to prove this? If your two locations already existed and had traditional phone systems, it would be a straightforward process. Take six months of voice bills from each location and look at calls between the facilities. Also examine calls to areas surrounding each location.

What if one of your locations is new and you don't have historical billing information for comparison? Contact the voice carriers and ask for their tariff tables (they all publish them). You can use these to estimate the cost of calls, giving you some idea of the costs and savings of using VoIP between the two locations.

Providing Support

If you are a multilocation company with limited staff, implementing VoIP may not necessarily increase the need for more full-time people, but it increases the need for support over the near term until everyone gets up to speed on the new way of doing telephony and using all the new and exciting features.

Choices for support are limited to two broad categories, each of which have their own peculiarities:

- Going in-house
- Using a VoIP partner

People will not tolerate an unstable telephony system. In the real-time world of business, a stable voice communications system is essential. Before deploying your VoIP system, run the traditional voice system in parallel with the VoIP system to make sure your configuration and circuits are functioning properly. Be sure to run a pilot test using some key decision-makers in your company; this will help sell the new technology.

In-house

Depending on the size of your network, in-house support can be a bonus if you have a dedicated, knowledgeable staff. A small staff supporting a global infrastructure, however, may not be the most optimal situation. A small staff may be desirable from a design and management perspective, but when supporting a global infrastructure, employing a hybrid of in-house people and partners may be better. When you start to consider the language barriers and cultural differences inherent in any global solution, the right partners make all the difference.

Your in-house team needs to closely monitor and identify changes to voice and data lines. Voice is not as forgiving as data traffic, so it needs to be prioritized over data and other application traffic; close monitoring helps ensure that voice traffic is given the priority it requires. An unannounced change to a provider's infrastructure can degrade the quality of your voice calls, so your in-house team needs to monitor and identify if or when this occurs. End-to-end testing between locations with the proper tools exposes any problems in your circuits. If you choose your tools poorly or do reactive monitoring, you are at the mercy of the circuit providers.

Partnering

VoIP is a relatively new technology, so I lean toward partnering as a way of augmenting existing staff. Partnering also has a training benefit — as your staff works closely with an experienced implementation partner, they quickly gain knowledge that is beneficial to your organization. Finding the right technology partner is important. Be sure you check references of any partner being considered. You might want to request a site visit to a company where the prospective partner has worked.

To get you started, see Chapter 18 for a description of some of the top VoIP manufacturers. Also see Appendix A for a list of some of the top VoIP carrier service providers. Most also lease traditional transport carrier services and have established customer service departments.

A partner can be indispensable when moving from a traditional voice system to a VoIP solution, especially if this is your first implementation of the technology. If you are in a dynamic environment, however, outside partners may shock you with their invoices. Changes cost money, and the more changes your partner needs to make in your company, the higher the cost. If you are in a rapidly changing environment, try to keep maintenance and support in-house as much as possible.

Keeping Up with Technology

Keeping current with telephony technology changes requires the same dedication and consistency as with other technologies. Your staff should stay current by joining local user groups and attending system-specific conferences. If none exist in your area, start one or attend meetings in neighboring cities. Stay current through trade magazines. Schedule monthly meetings with your VoIP partner to review system configurations and discuss upcoming system changes and enhancements.

Develop a relation with the vendor and schedule regular meetings to discuss your current system needs and future concerns. Request a copy of their annual plan and their current strategic objectives. The more information you can obtain, the better. Keep an eye on trends in the industry. Remember that not all changes or upgrades may be necessary for your system.

Before implementing any changes to an operational system, you may want to implement it on a test system. If you don't have a test system in-house, you may want to set up a contract with someone who does. This approach not only helps you test changes but can also be used to train your staff.

Part IV
The Part of Tens

"It's all here, Warden. Routers, hubs, switches, all pieced together from scraps found in the machine shop. I guess the prospect of unregulated telecommunications was just too sweet to pass up."

In this part . . .

Have you ever talked to someone and asked him or her to "cut to the chase"? Did that person then turn around and say "well, let me give you ten good reasons . . . "? If so, chances are good that you just talked to someone who finished reading a *For Dummies* book.

Welcome to the venerated Part of Tens — something you'll find in all *For Dummies* books. Here you find four chapters that give you ten good reasons for whatever is being discussed. In Chapter 15, you find ten good reasons for your company to switch to a VoIP system. Chapter 16 gives ten similar reasons for individuals to switch to VoIP.

In the adoption cycle of any new technology, rumor and innuendo swirl about. VoIP is no exception, so Chapter 17 explodes ten myths that may stop you from considering your own conversion. Finally, Chapter 18 introduces you to the top ten VoIP manufacturers — the go-to people in the industry.

All in all, the Part of Tens provides many more than ten reasons to sit up and pay attention; it's a great place to cut to the chase and quickly get just the information you need.

Chapter 15

Ten Reasons Why Your Company Should Switch to VoIP

In This Chapter

- Changing direction of the telephony industry
- Feature-rich, cost effective alternatives
- Existing investment protection
- Seamless maintenance and management
- Flexibility and portability
- Enhanced network management
- Better utilization of personnel
- Productivity applications
- Better bandwidth utilization
- Reduced costs

The reasons to switch to VoIP are countless, depending on how far you want to project the future of the marketplace. For now, here are the ten best reasons to make the switch.

Changing Direction of the Telephony Industry

Over the next few years, much of the $300 billion per year telecommunications industry will migrate and convert its equipment and carrier services to support packetized VoIP services on the WAN. It will not be long before traditional telephony systems providers are outdated.

As older providers lose customer base and revenue, they will streamline operations and eventually close their doors. The providers that stay in business will need to increase prices and therefore become noncompetitive. VoIP technology has become the strongest influence in the telecommunications provider marketplace.

As VoIP emerges worldwide as the number one replacement for traditional circuit-switched telephony infrastructure, manufacturers of telecommunications gear will convert their product lines to meet customer demands for VoIP-enabled systems. The same holds true for network services providers. They will convert their core service offerings to give priority to VoIP-related services. In fact, this is already occurring with most major carriers. The demand for circuit-switched equipment and network services will decline. As a result, the cost to suppliers who stay in the circuit-switched niche will go up. These costs will need to be passed on to the customers.

The leading VoIP and carrier services companies have made a commitment to developing secure and reliable IP telephony systems, communications software applications, life-cycle services, and carrier provider services. For a list of the leading manufacturers in the VoIP field, see Chapter 18. For a list of the leading VoIP providers, see Appendix A.

In light of newer VoIP products and services, customers will want to convert to VoIP so that they have adequate support available from outside companies. Many companies will also want to develop the VoIP skills of their in-house personnel. In this way, companies can insure their long-term growth by reducing costs and increasing revenue. VoIP can save companies lots of money in operating expenses, but if you have a multilocation company, converting to VoIP does require planning and VoIP skills.

Feature-Rich, Cost-Effective Alternatives

Most traditional telephony calling features have made their mark on the industry. Features such as voice mail, call transfer, call forwarding, and three-way calling have become familiar to all of us. The costs of these features are either rolled into the cost of your company's private telephony system, or you pay for them à la carte.

All traditional telephony features as well as many new features and communications applications are available in the brave new world of IP telephony. The number of calling features is overwhelming. And they all come with no additional cost because they are IP-based and are carried over the computer network.

Simple features such as being able to look at your telephony station and see a visual indicator that tells you whether someone in your calling group is "present" but on the telephone at the moment can help increase employee productivity. (Think how many times you wasted time calling people, only to get a busy signal or their voice mail, not knowing whether they were at their desk or not.) The presence feature is just one of many features available with VoIP.

Or how about the ability to run a soft phone on your computer and do telephony using a point-and-click process with a headset? Such a capability would never be contemplated in traditional telephony because that world can't support computer-related applications in a seamless manner. Many other calling features are available in VoIP, all just as compelling to companies considering a change.

Existing Investment Protection

If your company has a traditional telephone system (such as a PBX or KTS) in place, you can protect your investment by adapting the system in the new VoIP network. The PBX system probably includes many digital telephone stations. These telephones can also be reused in the new VoIP environment.

Your company can migrate to VoIP while protecting your existing telephony hardware investments.

A *forklift* upgrade is when you get rid of everything from the older system and therefore lose your previous investment. The other approach is to use some or all of your existing equipment. With the right VoIP partner, you can avoid forklift upgrades to VoIP.

Seamless Maintenance and Management

The full benefits of VoIP are realized in a *converged network* — one in which data and voice packets travel over the same infrastructure. Such a foundation eradicates redundant information systems, so the major tasks of installing and managing VoIP become more cohesive. Managers have more effective and direct applications to support their many challenges. They can manage not only computer data applications, but also IP-based telephony and videoconferencing systems. Unified database applications running over the network provide real-time, seamless access to all information needed to maintain the VoIP network.

Moves, adds, and changes that would require complex and costly resources in a traditional telephony network do not require the manager to do anything in a VoIP network. Instead, the network automatically adjusts itself to accommodate a user's new location. Usage, accounting, and other metrics data are available to the manager through any computer device attached to the network. With VoIP, managing and maintaining the network becomes cost-effective and seamless. Staff do not get caught up in problems and stay focused on business deliverables.

Flexibility and Portability

IP telephony has spawned many applications that increase both the flexibility and portability of communications. For instance, a soft phone provides mobile employees with easy access to real-time communications and the same calling features enjoyed by stationary employees. Users have never had more telephone options available for mobility. Wireless extension to cellular enables a "follow-me" feature so that employees can have calls ring at both their office and their cellular telephones.

In a VoIP network, employees can travel to any of the company's locations, plug in their IP-enabled laptop, begin work, and make and receive telephone calls. Employees have, at their distant temporary location, all the rich features normally available to them at their home office location. The network automatically identifies the user and applies that user's profile information. Employees can even direct their calls to any digital desktop telephone at the temporary location. (The telephone does not even have to be IP-enabled.) Managers no longer have to make costly and time-consuming accommodations for computer data and telephony connections for a coworker visiting their location.

Enhanced Network Management

VoIP provides a foundation for comprehensive network management. As a result, the ability for you to manage every bit and byte that runs over your LAN and WAN has never been more enabled.

Likewise, you have at your disposal tools that find and fix network issues so quickly that managers may rarely know that anything has happened. These types of tools can support local and remote network monitoring. In dedicated networks, near-perfect quality is provided. That's not to say that problems *never* occur, but with a VoIP network, your ability to detect symptoms and make changes to your setup in advance of any problems is greatly enhanced.

Better Utilization of Personnel

VoIP enables the realization of a converged network — data and telephony traveling over the same network. Gone are the days when you needed two different skill sets to maintain your networks (one for telephony and one for data). Although there are some skills unique to VoIP that traditional network engineers don't have, the underlying skills related to Ethernet networks and IP protocols are the same. This allows your company to maximize the training of your people and, in many cases, reduce the number of personnel you need in-house to support the network.

Productivity Applications

Many of the Web applications that previously ran exclusively over the Internet will now run over your private VoIP-based communications network. Your users can have their favorite Web page displayed on their VoIP telephone, or they can post special Web links on their telephone-based Web page. Many Web-based applications are candidates for running on your VoIP telephones.

Users can also add a video telephony solution, powered by IP video application software that enables a desktop PC or laptop to emulate an IP office phone. The quality of the video and audio that runs on the company's network, versus the Internet, is free from latency and jitter.

Better Bandwidth Utilization

Many people wrongly assume that when you add VoIP to an enterprise computer network, there won't be enough bandwidth available to support the change. The fact is that dedicated network transports supporting computer data or traditional telephony systems are about 30 percent utilized. Even though running both data and voice packets over the same network increases overall traffic, you must look at how the IP-based traffic operates.

On the LAN side, fault isolation provided by switching equipment maintains a steady mode of operation. If any chokepoints are identified, they can be remedied almost immediately by changing connection points or doing what the gurus call *load balancing.* But your IP-based management system will tell you this before it even becomes a problem.

On the WAN side, the load needs more consideration. You usually have more than one site on the WAN side that may have users connecting to your site. In addition, the cost and overall bandwidth capacity of WAN transports are higher and recur monthly when compared to the LAN side. (Chapter 7 provides more detail on transport lines and services to dynamically allocate bandwidth.)

For example, a T1 line has 24 channels. If you run traditional circuit-switched calls over the T1, you can maintain 24 simultaneous calls. The beauty of VoIP is that it is packetized, so the same 24 calls could run through just a fraction of the T1's overall capacity. As a result, you gain multiple times the bandwidth equivalent with VoIP when compared to circuit-switched telephony.

Reduced Costs

The cost reduction argument is compelling from a couple of perspectives. The argument is never more persuasive, however, than it is for companies that have a substantial volume of toll calls charged by the minute. VoIP can reduce local charges; that's a good thing. But VoIP also reduces or eliminates most other classes of toll charges and greatly reduces your regulatory fees. That is a great thing.

Depending on the number of locations your company has and how many toll boundaries your current calling plan covers, you can save big bucks. This savings is derived primarily from putting all your locations on VoIP and bypassing most if not all of your toll charges. If your organization has significant international calling, the same benefit accrues, except that your company can save even more on toll and regulatory costs. (See Chapter 3 for a breakdown on all five of the regulated carrier service categories.)

Chapter 16

Ten Reasons Why You Should Switch to VoIP at Home

In This Chapter

- One carrier
- One bill
- Free local service
- Reduced or eliminated toll service charges
- Reduced international toll charges
- More bandwidth
- Enhanced Internet access
- More ports to connect more phones and computers
- Wireless (WiFi) broadband service in your home
- Videoconferencing

Are you are a consumer? Do you have or want Internet access at home? If so, you may want to grab your calculator to add up all the things you can get with VoIP while reducing your monthly recurring charges for telephone and Internet access. The reasons to switch to VoIP at home are monthly savings, more services, piece of mind, and much more. In this chapter, I list the ten best reasons to make the switch.

One Carrier

How many times have you had to call your telephone carrier about one problem, your cable company about another problem, and your Internet access provider about yet another problem? Wouldn't it be nice to have just one provider with one telephone number?

One provider means that all your services are handled in the most expedient way. And you don't have to give up your POTS telephone service. You can continue to run your POTS line as part of your broadband services; in fact, you need to keep a POTS line if you're using DSL for your broadband access. Keeping your POTS line costs only about $25 per month, and your broadband provider may make you an even better deal if you bundle your POTS line together with your broadband Internet access. (See Chapter 6 for more information.)

One Bill

Working with one carrier means that you get one bill each month. If you have the DSL form of broadband, there is a surcharge for interstate carrier services that amounts to about 20 percent of your POTS line bill (approximately $5). Want to know why you need to pay the surcharge, even if you are using VoIP for your interstate calls? You'll need to ask your state and federal regulators; they are the ones who levy the surcharge. Unfortunately, we all have to pay the surcharge even though we may not use or even have a POTS line with interstate services on it.

Free Local Service

Even though you can't get away from some regulatory surcharges (see the preceding section), there is a way to offset the costs. Remember that you get free unlimited access to the local calling area over your POTS line. Therefore, if you use the POTS line for local calls and 911 calls, and VoIP for all other calling, you have no recurring telephony charges. The monthly savings, in most cases, more than offsets any surcharges you may need to pay.

Reduced or Eliminated Toll Service Charges

If you rack up a lot of minutes outside your local calling area, VoIP saves you a bundle of cash. (See Chapter 3 for the breakdown on different toll areas.) You can use your VoIP service to make all your far, far away calls and pay nothing for carrier charges.

Reduced International Toll Charges

If you make lots of international calls, VoIP is a way to greatly reduce the per-minute charges. The per-minute rates are reduced by a factor of ten or more compared to traditional carrier charges, depending on the country that you are calling. And you can call Canada for free with most VoIP plans.

More Bandwidth

If you are using a dialup modem to gain Internet access, you have one or more traditional POTS lines for telephony service. (You need a POTS line to hook up your modem.) The speed you get is excruciatingly slow compared to broadband. The bandwidth with broadband service is many times greater than with a dialup modem over a POTS line.

VoIP allows you to put all your telephony services on a single broadband account. You can keep one POTS line (your broadband carrier may provide one in your service bundle) for local calls and emergency purposes. You simply plug your POTS phone into the adapter box the carrier provides.

Enhanced Internet Access

Sure you get more bandwidth with broadband, but you also gain access to the entire Web. Everything from shopping on the Net and e-commerce to research and e-mail are now at your fingertips. Stay connected with your friends, your family, and your co-workers. Start a Web page; broadband providers often give customers Web site space for free. (Perhaps you can start a sideline business.)

Using VoIP, you can call people up wherever they may be located and talk as long as you want. You can even have a conference call and do instant messaging over the computer while the conference is going on. Broadband opens a whole new world for you; take advantage of it. Save money, gain services and features, and enjoy the good life.

More Ports to Connect More Phones and Computers

Getting broadband service does more than add VoIP telephony into your picture. It puts a piece of equipment (the cable or DSL modem) on your premises that lets you connect a multiport hub. Into this hub, you can plug several other devices (such as multiple computers or game machines), all of which can use the broadband line. One broadband line, many broadband services.

Wireless Broadband Service in Your Home

With broadband Internet access, you can plug in a wireless hub that gives you Internet and VoIP access without the need to run wires to your computer, provided your computer has wireless capability. With such capability, there is no longer a need to be tethered to the broadband box. If you have VoIP soft phone software, you can use the computer to connect to your wireless network and make VoIP calls.

Videoconferencing

Full-motion videoconferencing is now possible over broadband lines. Broadband connections provide enough bandwidth to support all your VoIP calling, your POTS line that runs over your broadband line, and your Internet access.

You can now add videoconferencing. You must purchase a video phone; the price range is $100 to $300. The phone plugs into your broadband provider's equipment and uses the same protocol software that VoIP uses.

Chapter 17

Ten VoIP Myths

In This Chapter

- VoIP runs only on the Internet
- POTS is cheaper
- POTS is faster
- The quality of service is suspect
- VoIP-enabled phones are pricey
- VoIP calls can be intercepted
- 911 calls may not work
- VoIP is not ready for prime time
- VoIP call features are expensive
- You have to throw out all your old PBX telephones

If a new technology comes our way that brings with it the promise of reducing or eliminating tremendous monthly costs, it can be expected that supporters and stakeholders of the status quo are going to be concerned. Consider the effect of the horseless carriage on commerce back in the early twentieth century. As a new market for automobiles emerged, many counter-arguments attempted to slow down its growth. But by 1910, there were an estimated five hundred thousand cars and the industry was growing fast. Stable owners, for example, became garage businesses. Saddle makers got into the business of providing seats for the new cars. It didn't take long for society to adopt the automobile and adapt to its presence.

In a similar way, VoIP is disrupting the $300 billion telecommunications market. VoIP has had to overcome many criticisms. This chapter addresses what many experts consider to be the most prevalent myths about VoIP.

VoIP Runs Only on the Internet

The most obvious myth about VoIP is that it runs only on the Internet. What can we expect? The term *Internet* is built into VoIP. However, VoIP requires and runs on the Internet protocol, but not necessarily on the Internet itself.

Internet protocol is not synonymous with the Internet. The Internet can be accessed from all types of networks. The Internet is not a network type unto itself; it is a network that is accessed by other networks and provides access to other networks.

VoIP runs on any network that can run the Internet protocol. This includes every network type known to man. (Chapter 4 discusses different network types.) But just because VoIP can run the same protocols over any network type, this does not mean VoIP runs the same way on all network types. The protocols take care of packetizing the telephony voice signals, but the network type takes care of transporting those packets.

POTS Is Cheaper

You might be inclined to say that VoIP is much more expensive than its older POTS counterpart. This comparison needs to be considered from several perspectives.

First you should think about the chronology. As with any technology, early pricing is never cheap. VoIP is about ten years old, and costs will continue to decrease.

In addition, if you price VoIP simply by the cost of the top-of-the-line hardware, you may draw a false conclusion. For example, a videophone that runs on VoIP is much more expensive than a low-end VoIP phone. But then again, full-motion, real-time videoconferencing would cost a lot less than flying to a distant location for a meeting.

Moreover, with VoIP, you don't need an entirely separate telephony system as you do with POTS-related telephony. VoIP substantially reduces the costs inherent to the traditional telephony infrastructure. For a large, global, multi-location company, these costs can easily soar into the millions of dollars.

Finally, look beyond the hardware and consider the services delivered. Check out your monthly telephone bill and try to understand exactly, line by line, what you are paying for. For companies with one or more locations, this is more complicated than it is for a consumer at home, but the process is the same — and just as illuminating.

POTS Is Faster

Some say POTS is faster than VoIP, but nothing could be further from the truth. When you evaluate the speed of VoIP, you need to consider what network or network types you plan to use to run VoIP. Private, dedicated networks running VoIP have proven to be as fast or even faster than POTS in supporting telephony. (See Chapter 7 for details.)

At the other end of the continuum, if you run VoIP as the very first Internet hobbyists did back in 1995, you are going to get slow speeds. The decisive factor here is that the public telephone network and the Internet sit between the VoIP caller and receiver. Because both ends used dialup modems and POTS lines for access, delay was inevitable.

The Quality of Service Is Suspect

The same arguments for speed can be made for QoS — specifically, what network type are we talking about? If, for example, you test the QoS of VoIP by limiting the network type to the circuit-switched PSTN, POTS QoS wins hands-down. After all, the PSTN is tuned for optimal POTS quality.

The major requirement with VoIP is that the voice signals get packetized. Consequently, the QoS with VoIP is determined by the network type. If you run your VoIP network over dedicated digital lines, you find that the QoS is just as good, if not better than, POTS QoS.

VoIP-Enabled Phones Are Pricey

Phones should be evaluated on the features and applications that they deliver to the customer. Taking this approach, some POTS phones cost more than VoIP phones, and vice versa. Moreover, many digital POTS phones work with a VoIP network. Home users are also using older analog phones (with a carrier-provided adapter) to run with VoIP over their broadband service.

With some VoIP phone types, you can do videoconferencing and Web surfing. You can even transform your computer into a telephone at no cost. Can a POTS telephone do all that?

VoIP Calls Can Be Intercepted

Can VoIP telephony packets on a computer network be intercepted? Yes, they can. What does it take to intercept VoIP packets? The same equipment and access that it takes to intercept computer data packets. How feasible is it? Not very.

After spending millions of dollars, the FBI developed a system called Carnivore that is essentially built on earlier network management technology known as a protocol analyzer. Basically, the device (a souped-up computer) plugs into a network much like any other network addressable device. It sits there and collects packets as they race by at the speed of light. The packets can then be analyzed for threats and other information, or so the theory goes.

If you're worried about such a device, keep the following in mind:

- A government agency at least as powerful as the FBI is required to gain access to a given network (excluding a trusted person doing it).
- Access must be physical. The person must have a key to the telecommunications closet or access to an office where they can plug into the network.
- Access is achieved through the network operating system, so the person must have a network access account.
- Network managers today have a variety of techniques to protect their packetized network traffic.

After all this, if you're still concerned about VoIP packet interception and security, consider the fact that anyone on the street can tap a POTS telephone line with a simple analog handset and a few wires. All they need is physical access to your line. They do not have to be inside your company; they can access the line from the street or in tunnels where the public access lines run.

A solid argument can be made that a packetized network has more security than the older circuit-switched network, particularly because you can also implement data encryption for VoIP.

911 Calls May Not Work

Remember that the 911 network was designed to be supported by and make full use of the circuit-switched network. VoIP uses packet-switched networks. This is a colossal difference that needs to be clearly understood.

At the same time, understand that most 911 calls are local calls to local emergency centers and law enforcement agencies. The major cost benefits associated with VoIP are not realized with local calls. Until the PSTN adopts VoIP and packet-switching, you have to maintain local POTS telephony service for local calls. Such lines can easily be used to make 911 calls directly. You don't have to lose your ability to make 911 calls just because you're converting to VoIP for all your toll calling.

On the consumer side, some low-end VoIP providers offer workarounds to enable their customers to let go of their POTS services and be fully VoIP. This is crazy. If you have an emergency, you want the fastest connection possible to 911. Why would you want a service that routes your emergency call out to the Internet, then to your VoIP provider, then back down to a POTS line, and finally to the local 911 center? My recommendation to consumers is to use DSL or cable modem for your Internet services; you can't get VoIP otherwise. Use the VoIP connection for all your toll calls and videoconferencing services. Plug a POTS telephone into your broadband VoIP adapter box and maintain at least one POTS line for local service.

Lastly, do not be fooled into thinking that you can use your cell phone to call 911 and thereby eliminate the need for that speedier $25 per month POTS line. Even a cell phone has a delay that is longer than what you would get from a standard POTS line. When you are mobile and on the road, you have no choice but to use your cell phone to contact 911. But in your home or business, there is no reason not to have a direct connection to your local 911.

VoIP Is Not Ready for Prime Time

We are beyond considering whether VoIP is here to stay when we can point to VoIP marketplace leaders (such as Avaya) that have a global market penetration base of more than a million companies and a Fortune 500 penetration of more than 90 percent. The problems that plagued VoIP in the 1990s have been overcome through technology or worked around with different technology.

VoIP is ready and working now.

VoIP Call Features Are Expensive

Because VoIP comes with all the usual call features you find in a traditional POTS call plan — plus several more features that are unavailable in the POTS world — many think that VoIP features add enormous costs to the monthly bill.

Adding features to your POTS line does add costs. But nothing could be further from reality when it comes to VoIP. VoIP operates with the TCP/IP protocols, which are used on the Web. As a result, VoIP features can be delivered through the software without additional costs. They are free in all private VoIP networks. In consumer VoIP service plans, the usual traditional features are included with the cost of the plan.

You Have to Throw Out All Your Old PBX Telephones

In the early days of VoIP, it quickly became apparent that the typical PBX (non-VoIP) environment could not run VoIP in its then-present state. Everything about the PBX was geared for circuit-switched telephony. The PBX back then could support dedicated access lines, but only if the lines were used to channelize the bandwidth for POTS line equivalencies. Because the PBX controlled all digital phones that came with the PBX, the conclusion back in the 1990s was that the PBX and its digital phones could not run VoIP.

Today, this potential drawback to VoIP has been wiped out. Through upgrades to the PBX, including connecting the PBX to the LAN, all digital phones that work with the PBX can now run VoIP. Companies today do not need to worry about losing their investments in PBX technology.

Chapter 18

Ten VoIP Manufacturers

In This Chapter

- Avaya
- Cisco
- Siemens
- Alcatel
- Nortel
- Mitel
- NEC
- 3COM
- Shoretel
- Inter-Tel

The VoIP telephony market is populated by many competitors, and you may need to choose among them to achieve the results you expect. To do IP telephony and VoIP correctly, you need to work with a manufacturer who can focus on using your existing resources to build a vibrant communications network that enhances productivity. Converged telephony systems should include all the traditional features you are familiar with (voice mail, call waiting, call forwarding, and so on) as well as many new exciting features that will really knock your socks off (such as presence and follow me).

Customers need reliability, and you want to work with a manufacturer that can deliver. If you are considering a converged network, you are probably anxious to gain a simple-to-manage, business-driven architecture at a competitive price. To accomplish your goals, you need a proven leader in this marketplace. This chapter provides a list of the best of the best.

In 2004, the METAspectrum for Enterprise IP Telephony report ranked the top IP telephony/VoIP companies. They ranked eight companies in the report (Avaya, Cisco, Siemens, Alcatel, Nortel, Mitel, NEC, and 3COM); each is included in this chapter. In addition, I rounded out the top ten with two other companies (Shoretel and Inter-Tel) that fared well in Gartner's 2004 report.

Avaya

Basking Ridge, New Jersey
866-462-8292
www.avaya.com

Avaya makes a wide variety of communications systems and software, including voice, converged voice and data, customer relationship management, messaging multiservice networking, and structured cabling products and services. According to Gartner, Avaya's "status as a leader is in part based on the architecture of its Avaya MultiVantage Communications Applications suite, which emphasizes an extensive feature set, scalability, consistent user interface, call processing power, and investment protection."

Avaya has a rich customer base with more than a million customers worldwide. Their products and systems are running in more than 90 percent of Fortune 500 companies. They are well known for their expertise in telephony systems, network integration, and unified communications management. Avaya is publicly held and has approximately fifteen thousand employees.

Cisco Systems

San Jose, California
800-553-6387
www.cisco.com

Cisco Systems makes networking solutions and network hardware and software, including converging voice and data products. According to Gartner, Cisco has "leveraged its strength in large-scale LAN infrastructure markets to win mind share among early adopters of converged networks. Its dealers are extremely effective in selling IT organizations, where many traditional telephony vendors are gaining credibility."

Cisco got its start in 1984 as a maker of networking products that could support proprietary and public data-networking protocols. For fiscal year 2004, they generated more than $24 billion in revenue. They are well known for their expertise in network routers and switches that provide the underlying framework for diverse technology networks.

Siemens

Munich, Germany
800-743-6367
`www.siemens.com`

Siemens is a publicly traded company that manufactures electronics and equipment for a range of industries, including information and communications, automation and control, power generation, transportation, medical, and lighting. They provide mobile communication and telephone communication systems to businesses and mobile phones and accessories to consumers. Siemens employs approximately seventy thousand people in the United States and four-hundred-thirty thousand worldwide, with global sales of more than $91 billion in 2004.

With respect to IP telephony, according to Gartner, "Siemens is focusing on its installed base of Hicom 300 systems, offering transition paths to IP for small, midsize and large businesses. However, migration gaps exist for users of pre-9006 release software, which Siemens is addressing with financial incentives designed to motivate customers to move to the HiPath 4000."

Alcatel

Paris, France
800-252-2835
`www.alcatel.com`

Alactel provides communications solutions to telecommunication carriers, Internet service providers, and enterprises. A publicly traded company with fifty-six thousand employees worldwide, Alcatel's focus is the delivery of voice, data, and video applications to customers and employees.

Their OmniPCX communications platform enables a company to selectively operate using traditional or IP telephony methods. The platform is capable of supporting hybrid operations as well.

Nortel

Brampton, Ontario, Canada
800-466-7835
www.nortel.com

Nortel was founded as Northern Electric in 1895. They make communications technologies and infrastructure equipment for service providers and enterprise customers, employing approximately thirty-five thousand people. Their revenues in 2004 were $9 billion.

According to Gartner, Nortel "has a broad IP telephony product portfolio that offers new prospects. Their Multimedia Communication Server 5100 offers presence management, collaboration tools, and a high degree of peer-to-peer communications throughout an organization."

Mitel

Kanata, Ontario, Canada
613-592-2122
www.mitel.com

Mitel is a maker of leading-edge business communications solutions for small- and medium-size organizations in more than fifty countries. Mitel is a privately held company employing more than two thousand people. They have gained much recognition in the IP telephony market because they focus on smaller companies with fewer than one hundred employees, and medium-sized companies with multiple locations that have fewer than two thousand employees per location.

NEC

Tokyo, Japan
212-326-2400
www.nec.com

Founded in 1905, NEC makes products ranging from computer hardware and software to wireless and IP telephony systems. For the fiscal year ending March 2005, NEC recorded more than $624 million in revenue. They employ about one-hundred-fifty thousand people worldwide.

According to Gartner, "NEC's portfolio offers various levels of converged IP capabilities, a multitude of features, scalability, and investment protection. Their platforms have an excellent reputation in the education, hospitality, and healthcare vertical markets, with attributes that can attract other organizations with distributed campus environments . . . NEC Unified Solutions strategy offers a menu of services that support the planning, implementation, network readiness and ongoing service needs of IP telephony."

3COM

Marlborough, Massachusetts
800-638-3266
`www.3com.com`

3COM has come a long way since the days when it was the world leader in the manufacture of network interface cards (NIC). Today, they continue to make networking and convergence products and services.

Founded in 1979, 3COM is a public company that employs approximately nineteen hundred people worldwide. Their NBX product line has met with great success in the small- to medium-sized marketplace. According to Gartner, "3COM enjoys an excellent reputation among users of its products, not only in IP telephony, but also data networking."

Shoretel

Sunnyvale, California
800-425-9385
`www.shoretel.com`

Shoretel, founded in 1998, is a privately held company that is all about IP telephony. Their approach is to evaluate your network first before designing a solution. The idea here is to determine how ready you are first, before taking the step into VoIP convergence.

According to Gartner, Shoretel's "product architecture gives organizations distributed call control across multiple locations through an IP backbone that supports the use of IP and analog telephones. This enables organizations to implement a converged network at their own pace."

Inter-Tel

Tempe, Arizona
480-449-8900
`www.inter-tel.com`

Inter-Tel was founded in 1969 and has grown from providing simple telephone systems for small businesses to providing sophisticated IP telephony and VoIP-based system solutions that connect multisite companies. They employ approximately nineteen hundred people.

According to Gartner, Inter-Tel's Axxess product line "provides cost-effective single-site solutions, as well as solutions for connecting multiple sites that can form a larger, single-system image. The platform supports voice and data convergence, networking, and call center and messaging applications."

Part V
Appendixes

"They both traveled a lot and were big Internet users. Finally, three years and two modems later, they broke up due to insufficient bandwidth."

In this part . . .

Unlike the human body (where some folks consider the appendix a useless part), the appendixes you find here actually provide some great information.

When it comes time to look for a competent VoIP partner, you'll find the information in Appendix A quite helpful. It ranks the top ten VoIP manufacturers, in order of revenue.

Are you befuddled by an unfamiliar word you've run across? Relax — chances are good that you can find the telecom or VoIP term you need in the glossary.

Appendix A

VoIP Providers

The VoIP provider market has been growing in leaps and bounds since AT&T announced in January 2004 that it was abandoning traditional carrier services as a way of doing business. What they were abandoning was the way of doing business outlined in the Telecommunications Act of 1996. The act defined ILECs, or Incumbent Local Exchange Carriers. ILECs (or simply LECs) owned all the physical lines in a given area. Everyone else had to go through ILEC to get any kind of access. The act also defined CLECs (Competitive Local Exchange Carriers). CLECs leased access lines at wholesale from the ILECs.

The customers, of course, ultimately paid the bill. Either they could get lines directly from an ILEC, or they could get them from a CLEC, who in turn got them from the same ILEC. Many analysts would say that the act had a beneficial effect on the marketplace, particularly with interstate long-distance costs. But in local markets, ILECs dominated because, no matter what, they had a piece of the action. And because ILEC controlled the actual physical installation of the access lines, any shortcomings in their service had a negative impact on the CLEC's customers.

CLECs, of which AT&T was one, are now using packet-switching protocols to provide VoIP services over their existing networks, rather than leasing POTS lines from ILECs at wholesale pricing. ILECs are now in a position where they could lose significant revenue. The fewer POTS lines installed by ILECs, the less revenue they have from wireline services.

For CLECs, the way is clear. Should they use circuit-switched POTS services at wholesale pricing, which they then mark up for their customers? Or do they use packet-switching services with VoIP, at a fraction of the wholesale price, using their own existing networks and offering their customers a much better value?

It is no small wonder why in the past six months the ILEC powers-that-be have sought to acquire the largest CLECs that offer VoIP. As a result of these changes in the marketplace, many argue that the corporate and consumer buying public will lose some of the choices they had when the lines between ILECs and CLECs were more clear. I disagree. With the inception of VoIP and the maturation of wireless, the choices for telephony and video services are increasing.

As a potential consumer, you need to know who is offering VoIP carrier services and who is offering the leased access lines you need to build or upgrade your VoIP network. Table A-1 lists the top ten VoIP carrier providers, ranked in order of gross revenue.

Table A-1 Top VoIP Carrier Providers

Rank	*Company*	*ILEC/CLEC*	*Revenue*	*Profit*	*Employees*
1	Verizon Communications	ILEC	$72B	$8B	212K
2	SBC Communications	ILEC	$41B	$6B	163K
3	AT&T	CLEC	$31B	($6B)	48K
4	Sprint	ILEC	$27B	($1B)	60K
5	BellSouth	ILEC	$23B	$5B	63K
6	MCI	CLEC	$23B	($4B)	40K
7	Comcast	CLEC	$20B	$1B	74K
8	Qwest Communications	ILEC	$14B	($2B)	41K
9	Nextel Communications	CLEC	$13B	$3B	19K
10	DIRECTV Group	CLEC	$12B	($2B)	12K

The first six carriers and number eight — Verizon, SBC, AT&T, Sprint, BellSouth, MCI, and Qwest — are able to lease or release any of the transports or transport services covered in Chapters 4 through 7. For example, to set up or upgrade a private dedicated network to run VoIP (see Chapter 7), your company most likely would work with one carrier to acquire the transports at each of your locations. The carriers can also offer VoIP services.

The seventh carrier, Comcast, is also the country's largest cable provider. The cable industry is not regulated, so the rules of the Telecommunication Act of 1996 do not apply to their cable services in the same way. They are permitted to be CLECs for the purpose of offering other services. For instance, if they lease cable service to your home, you may also get POTS service over that cable. For them to offer you POTS, they have to have transports into the PSTN. They can also lease VoIP services to you that run over your cable line.

The ninth carrier, Nextel Communications, is a cellular company, and the tenth, DIRECTV Group, is a satellite carrier company that offers broadband DSL to the consumer market.

The best place to start exploring for more information on any of the top ten is to visit their Web sites. Most of the sites have contact information and more details about their services:

- www.verizon.com
- www.sbc.com
- www.att.com
- www.sprint.com
- www.bellsouth.com
- www.mci.com
- www.comcast.com
- www.qwest.com
- www.nextel.com
- www.directv.com

Appendix B
Glossary

Numbers

911: The standard number for emergency telephone calls made over the public switched telephone network (PSTN) in North America. See also *E911.*

A

access: A generalized term referring to the physical means (such as POTS line or T1 line) through which a carrier company provides telecommunications services to a customer.

access cost: The one-time installation charges plus the recurring costs for access to one or more network services, such as Internet access.

add-on charges: Monthly charges beyond access costs. Add-on charges are mandated by the government and administered by the carrier. These types of add-on charges are also referred to as *regulatory fees.*

analog: A method of representing voice signals through a variation in the amplitude, frequency, or phase of an electrical signal. POTS telephone lines originally used (and in many areas continue to use) analog transmission to gain access to the public telephone network.

application layer: The top layer of the TCP/IP model. The application layer is where a packet is first encoded or last decoded. For the sender, it is the layer at which an application first presents information to the protocols to be packetized. For the receiver, it is the layer at which information is finally depacketized and ready for use by the application.

area code: A three-digit code that represents a specific geographic calling area in North America. The area code is dialed first in the sequence of calling any telephone number. Some areas, depending on population size, do not require an area code for calls made within the local regional calling area.

asymmetric transmission: When the upstream and downstream transmission rates are different; for example, the upload rate may be 256 Kbps and the download rate 1.536 Mbps. Most consumer broadband networks use asymmetric transmission.

B

B channel: A bandwidth unit employed by the integrated services digital network (ISDN). An ISDN B channel delivers 64 Kbps of digital bandwidth over the public switched telephone network (PSTN).

bandwidth: A measure of the amount of data that can be transmitted during a set period of time. In most cases, the terms *bandwidth* and *speed* are used synonymously. The bandwidth of a communications channel is often expressed in Kbps (kilobits per second) or Mbps (megabits per second).

bandwidth allocation: Digital bandwidth can be subdivided and allocated based on channels. For example, a T1 line with an aggregate bandwidth of 1.536 Mbps can allocate eight 64 Kbps channels to the telephone system, eight channels to the computer data network, and eight channels to a videoconferencing system. All 24 channels can be used in a dynamic allocation pool. A pool assigns channels as needed for the life of the telephone or videoconference call and returns the channels to the pool when the call or session is over.

basic rate interface: BRI. The consumer-grade level of ISDN service. A BRI consists of two B channels of 64 Kbps per channel. The two channels can be used in aggregate, or each channel can be dedicated to a specific application. A BRI also includes a D channel, which is a 16 Kbps channel used strictly by the carrier to manage services over the BRI line.

BRI: Basic rate interface. The consumer-grade level of ISDN service. A BRI consists of two B channels of 64 Kbps per channel. The two channels can be used in aggregate, or each channel can be dedicated to a specific application. A BRI also includes a D channel, which is a 16 Kbps channel used strictly by the carrier to manage services over the BRI line.

broadband: In general usage, a communications channel capable of transmission speeds equal to or greater than 256 Kbps. There is not currently a clear definition on the maximum speed for broadband. It also describes two popular services (DSL and cable modem) that connect consumers and small businesses to the Internet.

C

cable modem: A popular form of broadband service that runs over the consumer's cable television network transport line to offer access to the Internet.

call control: A PSTN management technology that establishes a connection, keeps the call up, and tears the call down when the parties hang up. Call control provides an automated means to track and manage call-related information for billing and maintenance.

call forwarding: A calling feature that enables the telephone customer to forward inbound telephone calls to another telephone number before ringing at the original destination number.

call transfer: A calling feature that enables the telephone customer to transfer an in-process telephone call to another telephone number.

calling feature: Additional uses or applications of the telephone, the telephone line, or the network that carries the telephone call. Voice mail, call forwarding, and call transfer are examples of traditional calling features. VoIP telephony has all the traditional calling features plus a new generation of features that use the telephone and the network, such as presence, vemail, and displaying a Web page on your telephone screen.

carrier: The company responsible for the transport lines used to provide communications services. Carriers lease transport lines to customers and often provide the services that operate over those lines, such as voice, data, and video transmission.

carrier services company: See *carrier.*

carrier services provider: See *carrier.*

carrier services infrastructure: CSI. Refers to a subset of five network types available in the telecommunications domain. Network transport lines for a particular CSI are available by leasing through carriers.

Centrex: Central Office Exchange. Centrex uses POTS lines to provide PBX-like services and features to customers.

channelization: The capacity for subdividing and allocating bandwidth channels in a dedicated transport line. For example, a DS3 transport line provides 672 channels of 64 Kbps each. Channelization allows those channels to be independently utilized for different communication purposes.

circuit-switched: The traditional method of transporting a telephone call over the PSTN. Multiple devices known as switches are employed by the carriers to form paths, or circuits, over which telephone calls may be carried between a caller and a receiver.

CLEC: Competitive local exchange carrier. Before the Telecommunications Act of 1996, the term interexchange carrier (IXC) was more commonly used. CLECs are the carriers that sought to lease transport lines and services to customers within an area where the Local Exchange Carriers (LECs) owned all the lines. The 1996 law mandated that the LECs (now ILECs) must lease their local lines at wholesale prices to CLECs. In return, the regulation promised to open the long-distance markets to ILECs. See also *ILEC.*

competitive local exchange carrier: CLEC. Before the Telecommunications Act of 1996, the term interexchange carrier (IXC) was more commonly used. CLECs are the carriers that sought to lease transport lines and services to customers within an area where the Local Exchange Carriers (LECs) owned all the lines. The 1996 law mandated that the LECs (now ILECs) must lease their local lines at wholesale prices to CLECs. In return, the regulation promised to open the long-distance markets to ILECs. See also *ILEC.*

compression: In digital networking, a method of reducing the bit length of network traffic (that is, packets) to enable more efficient data transmission.

convergence: The integration of switched and dedicated networks to support similar applications. For example, using VoIP on the corporate computer network to place a call on the traditional public telephone network.

CSI: Carrier service infrastructure. CSI refers to a subset of five network types available in the telecommunications domain. Network transport lines for a particular CSI are available by leasing through carriers.

D

D channel: A channel used by ISDN transport services. A BRI includes a 16-Kbps D channel, and a PRI includes a 64-Kbps D channel. These channels are used strictly by the carriers to manage services over the customer's ISDN line.

dedicated access: A classification of access using a private network transport (T1, T3, or OC3 line) through which a carrier provides ultra-high bandwidth telecommunications services to an individual customer.

dedicated network: A network dedicated to a single customer and implemented through the use of dedicated access lines.

delay: The total time it takes for a signal to travel from the source to the destination.

digital: A fast, efficient method of representing voice signals through high and low pulses. Since the inception of digital networking techniques, many newer, faster, and more precise methods of networking have become available.

digital service: DS. The initial form of the digital service CSI through which the first digital, private, dedicated transport lines (such as DS1 and DS3 lines) were installed beginning in 1964.

digital subscriber line: DSL. A popular form of broadband service which runs over a POTS line to provide Internet access. In most cases, DSL service requires the consumer to have an existing POTS line service.

digital telephony: Started first as a method for carriers to aggregate and transport POTS telephone calls on the carrier's network using DS-type transport lines. This same technology would later be redeveloped into privately owned telephone systems (PBX) that would be owned and operated by customers.

DS: See *digital service.*

DS0: One 64-Kbps channel of digital bandwidth on the DS network.

DS1: The standard for 24 DS0 channels having an aggregate bandwidth of 1.536 Mbps. A DS1 line is also known as a *T1 line.*

DS3: The standard for 672 DS0 channels having an aggregate bandwidth of 45 Mbps. A DS3 line is also known as a *T3 line.*

DSL: Digital subscriber line. A popular form of broadband service which runs over a POTS line to provide Internet access. In most cases, DSL service requires the consumer to have an existing POTS line service.

DWDM: Dense wave division multiplexing. A newer network transport service that aggregates network traffic for transmission in the terabit bandwidth ranges.

E

E911: An enhanced form of the standard number for emergency telephone calls made over the traditional PSTN in North America. E911 automatically provides, to the 911 call center agent, the caller's contact information, including name, address, and telephone number. The E911 system also has the capability to provide contact information to first responders, including police, fire, and paramedic personnel within the caller's local calling area.

encapsulation: A process whereby network traffic (data, voice, or video) is formatted according to the requirements of the network protocol (Ethernet or TCP/IP) being used to transport the traffic. The LAN encapsulates traffic into MAC frames. The WAN encapsulates traffic into packets.

encryption: A process that secures network traffic from unauthorized access. By using secure procedures or secure software keys known only to authorized users, the encrypted network traffic can be accessed.

Ethernet: The oldest and most popular protocol used for establishing data networks. Ethernet is used in more than 98 percent of corporate America for LAN networking. Ethernet is increasingly being used as a MAN backbone standard. The fundamentals of Ethernet are modified slightly to support WiFi and WiMax, popular forms of wireless Ethernet. See also *IEEE 802.3, IEEE 802.11,* and *IEEE 802.16.*

extensibility: An attribute of networks that allows them to be enlarged or enhanced without the need to change basic network characteristics.

F

feature: Additional uses or applications of the telephone, the telephone line, or the network that carries the telephone call. Voice mail, call forwarding, and call transfer are examples of traditional calling features. VoIP telephony has all the traditional calling features plus a new generation of features that use the telephone and the network, such as presence, vemail, and displaying a Web page on your telephone screen.

feature cost: Charges associated with the implementation of calling features. In traditional POTS telephony, calling features have a cost associated with them, typically in the form of additional monthly charges.

fiber optic: A physical cable medium used in most networks for all outside segments of the network's physical layer.

firewall: Software or hardware that limits access to a data network. Some firewall systems also provide network management functions.

frame: On the LAN side of the network, bit signal traffic is encapsulated and transported inside MAC frames. See also *MAC.*

G

gateway: A network device used to provide access between different types of networks. For instance, a gateway may provide access into an external network such as the PSTN, the Internet, or a private WAN. A PSTN gateway has a LAN interface on the inside and a PRI access transport line on the outside. It translates IP telephony frames from the LAN into circuit-switched POTS traffic for the PSTN and vice versa.

H

hard phone: A VoIP-enabled telephone that has an RJ-45 LAN interface port to connect it to the Ethernet LAN. VoIP telephones today come in all shapes and sizes.

HFC: Hybrid fiber-coaxial. The CSI that supports cable television, broadband services, and VoIP telephony through cable modem and POTS telephony through a POTS telephone and adapter.

hop: On the Internet, a hop represents a single, intermediary step in the path of a network transmission from the source to the destination.

hosted VoIP: A managed VoIP telephony service similar in concept to the traditional Centrex model. Synonymous with *IP Centrex, VoIP Centrex, hosted telephony,* and *hosted VoIP telephony.*

hybrid fiber-coaxial: HFC. The CSI that supports cable television, broadband services, and VoIP telephony through cable modem and POTS telephony through a POTS telephone and adapter.

I

IEEE: Institute for Electrical and Electronic Engineers. The main standards-certifying body for protocols such as Ethernet (IEEE 802.3), WiFi (IEEE 802.11), and WiMax (IEEE 802.16).

IEEE 802.3: The standard defining early forms of the Ethernet networking protocol.

IEEE 802.11: The standard defining early forms of wireless Ethernet (WiFi).

IEEE 802.16: The standard defining a form of WiFi known as WiMax designed to handle higher bandwidths over greater distances.

ILEC: Incumbent local exchange carrier. Introduced with the Telecommunications Act of 1996. ILEC is intended to identify the carrier who owns the traditional, regulated cabling infrastructure in any given LATA. See also *CLEC.*

in-state toll: One of four traditional regulated toll carrier service categories, also known as *intrastate.*

integrated services digital network: ISDN. A group of digital transport services that use the circuit-switched PSTN. ISDN transports are capable of integrating data, voice, and video applications, but run slower than other transports available today.

interexchange carrier: IXC. Dominated the long-distance carrier services marketplace before the implementation of the Telecommunications Act of 1996.

international: One of four traditional regulated toll carrier service categories. Considered to be the most highly regulated of all toll services.

Internet: A global, publicly accessible, nonregulated, nonsecure network accessible from all five CSIs.

Internet service provider: ISP. A company that provides Internet access to consumers and companies. Larger and more versatile ISPs offer Internet access using a variety of network transport options.

interstate toll: One of four traditional regulated toll carrier service categories, also known as *long distance.*

intralata toll: One of four traditional regulated toll carrier service categories, also known as *local toll* or *regional toll.* Refers to calls in which the caller and receiver are in the same local access and transport area (LATA).

intranet: A private network based on the same protocols used on the larger public Internet. Those outside the network can gain access to a corporate intranet through a firewall or gateway, if the network is configured to allow such access.

intrastate toll: One of four traditional regulated toll carrier service categories, also known as *in-state toll.*

IP: Internet Protocol. One of two major protocols used in the TCP/IP family of protocols. The IP protocol is one of the protocols used to implement the Internet.

IP address: An address comprised of four numbers, each ranging from 0 to 255, and normally expressed with each number separated by a period (such as 192.168.2.100). IP addresses are used to route network traffic from sender to receiver. The IP address is a major component field of a VoIP packet and is used to map the VoIP telephone call to a specific telephone number. In a VoIP telephony call, both source and destination (caller and receiver) addresses are used to establish and maintain the VoIP call.

IP Centrex: A managed VoIP telephony service similar in concept to the traditional Centrex model. Synonymous with *VoIP Centrex, hosted VoIP, hosted telephony,* and *hosted VoIP telephony.*

IP soft phone: Software that enables a computer to function as a VoIP telephone, including an on-screen dialing pad for point-and-click dialing.

IP telephony: IPT. A technology that allows traditional voice calls to be carried as data over a local area network. IPT is technically VoIP on a LAN (and VoIP is IPT outside the LAN).

IPT: IP telephony. A technology that allows traditional voice calls to be carried as data over a local area network. IPT is technically VoIP on a LAN (and VoIP is IPT outside the LAN).

ISDN: Integrated services digital network. A group of digital transport services that use the circuit-switched PSTN. ISDN transports are capable of integrating data, voice, and video applications, but run slower than other transports available today.

ISP: Internet service provider. A company that provides Internet access to consumers and companies. Larger and more versatile ISPs offer Internet access using a variety of network transport options.

IXC: Interexchange carrier. IXCs dominated the long-distance carrier services marketplace before the implementation of the Telecommunications Act of 1996.

K

key telephone system: KTS. An internal phone system typically used by smaller companies so they can take advantage of calling features and minimize the number of POTS lines necessary. A KTS is often referred to as a key station or simply a key system.

KTS: See *key telephone system*.

L

LAN: Local area network. A data network limited to a small geographic area. A LAN can be as small as a couple of devices connected on the same network or as large as a campuswide installation with numerous buildings and thousands of addressable devices on the same network.

last mile: The physical line installed by the carrier to support the connection between the customer's premise and the carrier's point-of-presence. The local loop, sometimes referred to as the *local loop,* is used to enable access to one or more networks and carrier services.

LATA: Local access and transport area. An arbitrary geographical regulatory designation established by ILECs.

LEC: See *ILEC.*

line: A physical channel (includes wireless channels) or aggregate of contiguous channels that supports the transmission of electrical, optical, or telemetric data, voice, or video bit-level signaling. Also known as *transport.*

line cost: The initial start-up installation charges, plus the recurring costs, for a network transport line to connect the customer's premise to the carrier's point-of-presence.

local access and transport area: LATA. An arbitrary geographical regulatory designation established by ILECs.

local area network: LAN. A data network limited to a small geographic area. A LAN can be as small as a couple of devices connected on the same network or as large as a campuswide installation with numerous buildings and thousands of addressable devices on the same network.

local calling area: A regulated calling area usually covering the immediate surrounding geographical area. Distinguishes local calling from all four toll related calling service categories.

local exchange carrier: See *ILEC.*

local loop: The physical line installed by the carrier to support the connection between the customer's premise and the carrier's point-of-presence. The local loop, sometimes referred to as the *last mile,* is used to enable access to one or more networks and carrier services.

local toll: One of four traditional regulated toll carrier service categories, also known as *intralata toll* or *regional toll.* Refers to calls in which the caller and receiver are in the same local access and transport area (LATA).

long distance: One of four traditional regulated toll carrier service categories, also known as *interstate toll.*

M

MAC: (1) In network terminology, an acronym for media access control. The part of the network interface that controls physical access to the LAN through the MAC address. (2) In telephone system administration terminology, an acronym for moves, adds, and changes. MAC describes the most common type of maintenance necessary in traditional telephone systems.

MAC address: An address that uniquely identifies a network device. The MAC address is typically represented in hexadecimal notation, as in 00-04-23-58-90-6E.

MAC frame: The format specified for encapsulating bit signals in the IEEE 802.3, IEEE 802.11, and IEEE 802.16 standards.

MAN: Metropolitan area network. A type of network designed to cover a large geographical area, such as a city.

media access control: MAC. The part of the network interface that controls physical access to the LAN through the MAC address.

metropolitan area network: MAN. A type of network designed to cover a large geographical area, such as a city.

monthly recurring cost: MRC. In the regulated communications industry, charges that are assessed monthly for access, calling features, tolls, and regulatory fees.

moves, adds, and changes: MAC. The most common type of maintenance necessary in traditional telephone systems.

MRC: Monthly recurring cost. In the regulated communications industry, charges that are assessed monthly for access, calling features, tolls, and regulatory fees.

N

network feature: Calling features provided in a VoIP network. For example, presence, Web surfing, and vemail are network features.

network interface card: NIC. Provides the network device, such as a computer or a VoIP telephone, with its MAC address and the means for connecting to the LAN.

network management system: NMS. Provides bit-level metrics and statistics on network utilization, error rates, bottlenecks, security breeches, frame faults, and faulty packets.

network timing protocol: NTP. Enables timing so that the composure of the voice signals (high and low pulses that make up the voice pattern) being sent is the same relative composure of the voice signals received. Unnecessary delay or not enough time allowance causes variation and possibly distortion.

NIC: Network interface card. Provides the network device, such as a computer or a VoIP telephone, with its MAC address and the means for connecting to the LAN.

NMS: Network management system. Provides bit-level metrics and statistics on network utilization, error rates, bottlenecks, security breeches, frame faults, and faulty packets.

NPA: Numbering plan area. Usually corresponds to the area code in the traditional POTS-PSTN telephone numbering system.

NTP: Network timing protocol. Enables timing so that the composure of the voice signals (high and low pulses that make up the voice pattern) being sent is the same relative composure of the voice signals received. Unnecessary delay or not enough time allowance causes variation and possibly distortion.

number exchange: NXX. Usually corresponds to the prefix in the traditional POTS-PSTN telephone numbering system.

numbering plan area: NPA. Usually corresponds to the area code in the traditional POTS-PSTN telephone numbering system.

NXX: Number exchange. Usually corresponds to the prefix in the traditional POTS-PSTN telephone numbering system.

O

OC network: Optical carrier network. One of the five CSIs, the OC network is implemented through the use of fiber-optic cabling and extremely high bandwidth data transfers.

OC12: An OC CSI transport that provides 622 Mbps of digital bandwidth.

OC3: An OC CSI transport that provides 155 Mbps of digital bandwidth.

off-net: In VoIP telephony, refers to calls that must be carried on another network (usually the PSTN) external to the VoIP network.

on-net: In VoIP telephony, refers to calls carried on the customer's network.

optical carrier network: OC network. One of the five CSIs, the OC network is implemented through the use of fiber-optic cabling and extremely high bandwidth data transfers.

P

packet: On the WAN side of the network, bit-signal traffic is encapsulated and transported inside packets. A packet can be best visualized as an electronic envelope for transmitting data.

packet-switched: Packet-switched networks such as a VoIP network use the addressing information contained in the packet to determine the route the packet takes to its destination.

PBX: Private branch exchange. A telephone system used by larger companies to manage POTS-PSTN telephony services and calling features.

physical layer: In the TCP/IP networking model, the physical layer is where all packets are converted to electro or electro-optical signals to be carried over the local or external network.

plain old telephone service: POTS. The most basic form of circuit-switched telephone service.

point of presence: POP. A brick-and-mortar facility where a CLEC has established an operating presence in the local carrier exchange marketplace. In a given LATA, many CLECs may have one or more POPs. Under regulations introduced in 1996, the CLECs can acquire transport lines and access to connect their customer's premises through the ILEC's network to the CLEC's POP(s).

POP: Point of presence. A brick-and-mortar facility where a CLEC has established an operating presence in the local carrier exchange marketplace. In a given LATA, many CLECs may have one or more POPs. Under regulations introduced in 1996, the CLECs can acquire transport lines and access to connect their customer's premises through the ILEC's network to the CLEC's POP(s).

POTS: Plain old telephone service. The most basic form of circuit-switched telephone service.

POTS line: The physical line that supports plain old telephone service (POTS).

POTS telephone: The telephone that supports plain old telephone service (POTS).

presence: One of the network features available with VoIP telephony services through a light indicator or software icon. If the presence light is on, the person is available on the network.

PRI: Primary rate interface. An ISDN transport line providing 23 B channels and one 64-Kbps D channel. The PRI has gained renewed popularity with the advent of VoIP networks that use gateways with PRIs to support off-net calls to the PSTN.

primary rate interface: PRI. An ISDN transport line providing 23 B channels and one 64-Kbps D channel. The PRI has gained renewed popularity with the advent of VoIP networks that use gateways with PRIs to support off-net calls to the PSTN.

private branch exchange: PBX. A telephone system used by larger companies to manage POTS-PSTN telephony services and calling features.

PSTN: Public switched telephone network. The oldest and largest communications network in the world.

PSTN baseline: Because of its history and high quality of service, VoIP technology uses the PSTN as a baseline for developing and designing telephony networks based on VoIP.

PSTN gateway: A gateway is a network device used to provide access between different types of networks. For instance, a gateway may provide access into an external network such as the PSTN, the Internet, or a private WAN. A PSTN gateway has a LAN interface on the inside and a PRI access transport line on the outside. It translates IP telephony frames from the LAN into circuit-switched POTS traffic for the PSTN and vice versa.

public switched telephone network: PSTN. The oldest and largest communications network in the world.

R

real-time transport control protocol: RTCP. Operates at the application layer of the TCP/IP model to monitor voice signal delivery and provide minimal control functions to ensure delivery of packets.

real-time transport protocol: RTP. Operates at the application layer of the TCP/IP model to provide end-to-end network transport functions for digital voice signals encapsulated in the VoIP packet.

regional toll: One of four traditional regulated toll carrier service categories, also known as *intralata toll* or *local toll.* Refers to calls in which the caller and receiver are in the same local access and transport area (LATA).

regulatory fees: Add-on charges to telephone bills using POTS-PSTN access lines. Whether you use the services or not, having the POTS access line is sufficient for being charged most regulatory fees.

router: A network device that connects the LAN to one or more external networks. The router translates frame traffic on the LAN into packetized traffic for the WAN or the Internet.

RTCP: Real-time transport control protocol. Operates at the application layer of the TCP/IP model to monitor voice signal delivery and provide minimal control functions to ensure delivery of packets.

RTP: Real-time transport protocol. Operates at the application layer of the TCP/IP model to provide end-to-end network transport functions for digital voice signals encapsulated in the VoIP packet.

S

scalability: An attribute of networks that allows them to increase capacity without the need to change basic network characteristics.

session initiation protocol: SIP. An interoperable protocol in the TCP/IP family of protocols. SIP uses text formatting to set up and maintain communication sessions with various endpoints. These endpoints can include cell phones, desk phones, PC clients, and PDAs. SIP permits these various endpoints to operate as a single system.

signaling system: SS7. A trunking protocol used in the PSTN to control POTS telephone calls including channel setup, session maintenance, and call tear down.

SIP: Session initiation protocol. An interoperable protocol in the TCP/IP family of protocols. SIP uses text formatting to set up and maintain communication sessions with various endpoints. These endpoints can include cell phones, desk phones, PC clients, and PDAs. SIP permits these various endpoints to operate as a single system.

soft phone: Software that enables a computer to function as a VoIP telephone, including an on-screen dialing pad for point-and-click dialing.

splitter: A device that terminates the cable from the cable company and divides the signal carried over that cable into component services, such as cable television, Internet access, VoIP telephony, and POTS telephony.

SS7: Signaling system 7. A trunking protocol used in the PSTN to control POTS telephone calls including channel setup, session maintenance, and call tear down.

symmetric transmission: When the upstream and downstream transmission rates on a broadband connection are the same.

T

T1: The standard for 24 DS0 channels having an aggregate bandwidth of 1.536 Mbps. A T1 line is also commonly known as a *DS1 line.*

T3: The standard for 672 DS0 channels having an aggregate bandwidth of 45 Mbps. A T3 line is also known as a *DS3 line.*

TCP: Transmission control protocol. One of two major protocols used in the TCP/IP family of protocols. In VoIP telephony and videoconferencing calls, the TCP protocol is replaced by its sister protocol, UDP.

TCP/IP: Transmission control protocol/Internet protocol. The family of interoperable protocols consisting of more than one-hundred-twenty protocols, each of which performs one or more services to support various network applications. The early developers of the Internet agreed upon the name TCP/IP because, at the time, TCP and IP were considered the two most important protocols for any network connection.

Telecommunications Act of 1996: Extensive legislation that marked a turning point in the efforts to promote competition in the former local and long-distance carrier services marketplace.

toll bypass: A term that concisely describes how VoIP telephony service completely sidesteps the regulated, circuit-switched PSTN and all its associated toll usage charges by carrying telephone calls over private, packet-switched networks.

toll charges: Recurring, metered, per-minute charges assessed on any POTS-PSTN call that terminates in any of the four toll service category zones.

transmission control protocol: TCP. One of two major protocols used in the TCP/IP family of protocols. In VoIP telephony and videoconferencing calls, the TCP protocol is replaced by its sister protocol, UDP.

transmission control protocol/Internet protocol: TCP/IP. The family of interoperable protocols consisting of more than one-hundred-twenty protocols, each of which performs one or more services to support various network applications. The early developers of the Internet agreed upon the name TCP/IP because, at the time, TCP and IP were considered the two most important protocols for any network connection.

transport: A physical channel (includes wireless channels) or aggregate of contiguous channels that supports the transmission of electrical, optical, or telemetric data, voice, or video bit-level signaling. In common usage, *transport* is often used synonymously with *line*.

transport layer: In the TCP/IP model of networking, the layer at which the end-to-end transport of frames or packets occurs.

transport service: An application supporting the transmission of data, voice, or video signals, or any combination of these signals, over a given transport, according to a specific protocol or set of protocols.

triple play: Refers to the capability to integrate data, voice, and video applications on the same transport.

U

UDP: User datagram protocol, an encoding protocol implemented at the transport layer of VoIP telephony and videoconferencing calls.

V

vemail: A network feature supported with VoIP telephony in which the user can elect to hear their e-mail or print a hardcopy of their voice mail.

virtual private network: VPN. A network that interconnects multiple local area networks (LANs) by using the Internet as a WAN transport or backbone.

voice mail: A popular calling feature that allows callers to leave a message in the event that the called party can't answer the call. Voice mail comes at no additional cost with VoIP telephony.

voice over Internet protocol: VoIP. A network service that supports carrying telephone calls over packetized networks. VoIP reduces substantially or eliminates the need for a separate, circuit-switched telephone network to carry telephone calls.

VoIP: Voice over Internet protocol. A network service that supports carrying telephone calls over packetized networks. VoIP reduces substantially or eliminates the need for a separate, circuit-switched telephone network to carry telephone calls.

VoIP adapter box: In broadband services, an adapter box is provided by the carrier to enable VoIP customers to plug in a VoIP telephone or to continue to use their existing POTS telephones, if desired.

VoIP Centrex: A managed VoIP telephony service similar in concept to the traditional Centrex model. Synonymous with *IP Centrex, hosted telephony, hosted VoIP telephony,* and *hosted VoIP.*

VPN: Virtual private network. A network that interconnects multiple local area networks (LANs) by using the Internet as a WAN transport or backbone.

W

WAN: Wide area network. A larger network that connects two or more LANs using dedicated transport lines.

WAP: Wireless access point. A device that enables network devices with wireless interface cards to connect to the network. Wireless users (WiFi or WiMax) must be within the specified distance range limits. The WAP has one cable, which is used to connect it to the larger hardwired network.

WEC: Wireless extension to cellular. A technology that permits a user to have their desk phone ring on their cellular telephone and not be charged as a cellular call. Essentially, the company's network is used to forward the call as if the cell phone were another station on the network.

wide area network: WAN. A larger network that connects two or more LANs using dedicated transport lines.

WiFi: Wireless fidelity. Wireless networking as specified in the IEEE 802.11 series of standards.

WiMax: Worldwide interoperability for microwave access. Wireless networking as defined in the IEEE 802.16 standard. WiMax is demonstrating speeds in excess of 70 Mbps (more than six times the maximum speed of WiFi) over much greater distances (up to 30 miles).

wireless access point: WAP. A device that enables network devices with wireless interface cards to connect to the network. Wireless users (WiFi or WiMax) must be within the specified distance range limits. The WAP has one cable, which is used to connect it to the larger hardwired network.

wireless extension to cellular: WEC. A technology that permits a user to have their desk phone ring on their cellular telephone and not be charged as a cellular call. Essentially, the company's network is used to forward the call as if the cell phone were another station on the network.

wireless fidelity: WiFi. Wireless networking as specified in the IEEE 802.11 series of standards.

worldwide Interoperability for microwave access: WiMax. Wireless networking as defined in the IEEE 802.16 standard. WiMax is demonstrating speeds in excess of 70 Mbps (more than six times the maximum speed of WiFi) over much greater distances (up to 30 miles).

Index

• C •

• E •

• F •

• G •

• H •

• I •

• J •

• K •

• L •

• M •

• N •

• O •

• P •

• Q •

• R •

• S •

• T •

• U •

• V •

• W •

Notes

Notes

Notes

Notes

Notes

Notes

Notes

SPORTS, FITNESS, PARENTING, RELIGION & SPIRITUALITY

0-7645-5146-9

0-7645-5418-2

Also available:

- Adoption For Dummies 0-7645-5488-3
- Basketball For Dummies 0-7645-5248-1
- The Bible For Dummies 0-7645-5296-1
- Buddhism For Dummies 0-7645-5359-3
- Catholicism For Dummies 0-7645-5391-7
- Hockey For Dummies 0-7645-5228-7
- Judaism For Dummies 0-7645-5299-6
- Martial Arts For Dummies 0-7645-5358-5
- Pilates For Dummies 0-7645-5397-6
- Religion For Dummies 0-7645-5264-3
- Teaching Kids to Read For Dummies 0-7645-4043-2
- Weight Training For Dummies 0-7645-5168-X
- Yoga For Dummies 0-7645-5117-5

TRAVEL

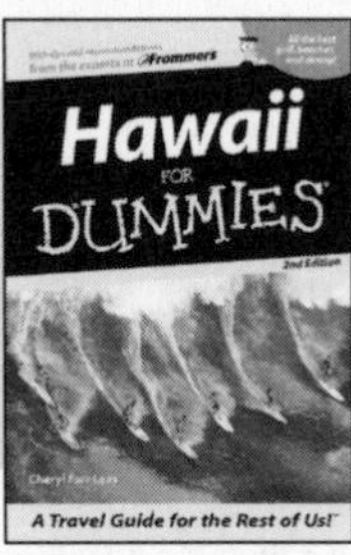

0-7645-5438-7

0-7645-5453-0

Also available:

- Alaska For Dummies 0-7645-1761-9
- Arizona For Dummies 0-7645-6938-4
- Cancún and the Yucatán For Dummies 0-7645-2437-2
- Cruise Vacations For Dummies 0-7645-6941-4
- Europe For Dummies 0-7645-5456-5
- Ireland For Dummies 0-7645-5455-7
- Las Vegas For Dummies 0-7645-5448-4
- London For Dummies 0-7645-4277-X
- New York City For Dummies 0-7645-6945-7
- Paris For Dummies 0-7645-5494-8
- RV Vacations For Dummies 0-7645-5443-3
- Walt Disney World & Orlando For Dummies 0-7645-6943-0

GRAPHICS, DESIGN & WEB DEVELOPMENT

0-7645-4345-8

0-7645-5589-8

Also available:

- Adobe Acrobat 6 PDF For Dummies 0-7645-3760-1
- Building a Web Site For Dummies 0-7645-7144-3
- Dreamweaver MX 2004 For Dummies 0-7645-4342-3
- FrontPage 2003 For Dummies 0-7645-3882-9
- HTML 4 For Dummies 0-7645-1995-6
- Illustrator CS For Dummies 0-7645-4084-X
- Macromedia Flash MX 2004 For Dummies 0-7645-4358-X
- Photoshop 7 All-in-One Desk Reference For Dummies 0-7645-1667-1
- Photoshop CS Timesaving Techniques For Dummies 0-7645-6782-9
- PHP 5 For Dummies 0-7645-4166-8
- PowerPoint 2003 For Dummies 0-7645-3908-6
- QuarkXPress 6 For Dummies 0-7645-2593-X

NETWORKING, SECURITY, PROGRAMMING & DATABASES

0-7645-6852-3

0-7645-5784-X

Also available:

- A+ Certification For Dummies 0-7645-4187-0
- Access 2003 All-in-One Desk Reference For Dummies 0-7645-3988-4
- Beginning Programming For Dummies 0-7645-4997-9
- C For Dummies 0-7645-7068-4
- Firewalls For Dummies 0-7645-4048-3
- Home Networking For Dummies 0-7645-42796
- Network Security For Dummies 0-7645-1679-5
- Networking For Dummies 0-7645-1677-9
- TCP/IP For Dummies 0-7645-1760-0
- VBA For Dummies 0-7645-3989-2
- Wireless All In-One Desk Reference For Dummies 0-7645-7496-5
- Wireless Home Networking For Dummies 0-7645-3910-8

HEALTH & SELF-HELP

0-7645-6820-5 *†

0-7645-2566-2

Also available:

- Alzheimer's For Dummies 0-7645-3899-3
- Asthma For Dummies 0-7645-4233-8
- Controlling Cholesterol For Dummies 0-7645-5440-9
- Depression For Dummies 0-7645-3900-0
- Dieting For Dummies 0-7645-4149-8
- Fertility For Dummies 0-7645-2549-2
- Fibromyalgia For Dummies 0-7645-5441-7
- Improving Your Memory For Dummies 0-7645-5435-2
- Pregnancy For Dummies † 0-7645-4483-7
- Quitting Smoking For Dummies 0-7645-2629-4
- Relationships For Dummies 0-7645-5384-4
- Thyroid For Dummies 0-7645-5385-2

EDUCATION, HISTORY, REFERENCE & TEST PREPARATION

0-7645-5194-9

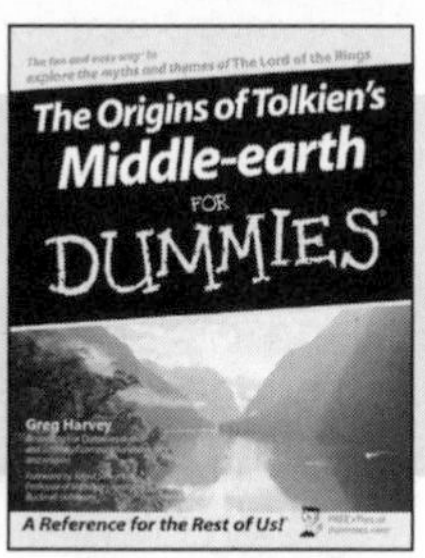

0-7645-4186-2

Also available:

- Algebra For Dummies 0-7645-5325-9
- British History For Dummies 0-7645-7021-8
- Calculus For Dummies 0-7645-2498-4
- English Grammar For Dummies 0-7645-5322-4
- Forensics For Dummies 0-7645-5580-4
- The GMAT For Dummies 0-7645-5251-1
- Inglés Para Dummies 0-7645-5427-1
- Italian For Dummies 0-7645-5196-5
- Latin For Dummies 0-7645-5431-X
- Lewis & Clark For Dummies 0-7645-2545-X
- Research Papers For Dummies 0-7645-5426-3
- The SAT I For Dummies 0-7645-7193-1
- Science Fair Projects For Dummies 0-7645-5460-3
- U.S. History For Dummies 0-7645-5249-X

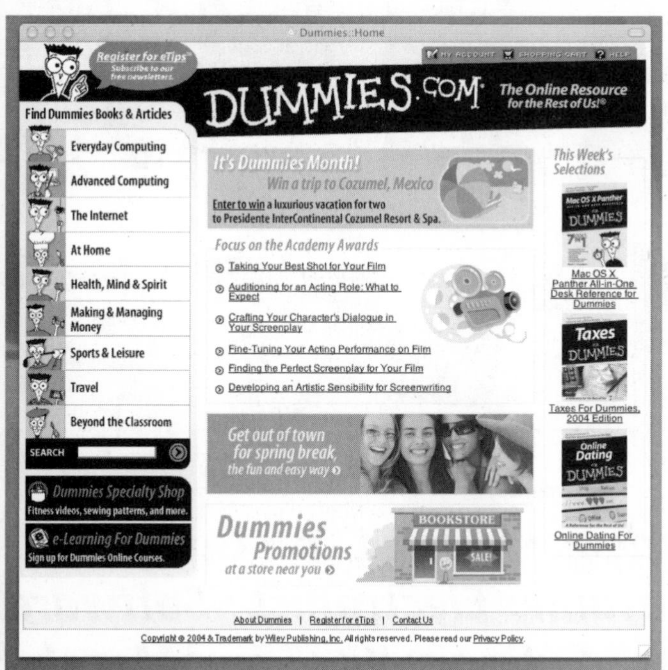

Get smart @ dummies.com®

- **Find a full list of Dummies titles**
- **Look into loads of FREE on-site articles**
- **Sign up for FREE eTips e-mailed to you weekly**
- **See what other products carry the Dummies name**
- **Shop directly from the Dummies bookstore**
- **Enter to win new prizes every month!**

*** Separate Canadian edition also available**
† Separate U.K. edition also available

Available wherever books are sold. For more information or to order direct: U.S. customers visit www.dummies.com or call 1-877-762-2974. U.K. customers visit www.wileyeurope.com or call 0800 243407. Canadian customers visit www.wiley.ca or call 1-800-567-4797.